PHOTOVOLTAIC ENERGY SYSTEMS

Matthew Buresch

PHOTOVOLTAIC ENERGY SYSTEMS

Design and Installation

McGraw-Hill Book Company

New York St. Louis San Francisco Auckland
Bogotá Hamburg Johannesburg London Madrid
Mexico Montreal New Delhi Panama Paris
São Paulo Singapore Sydney Tokyo Toronto

Library of Congress Cataloging in Publication Data

Buresch, Matthew.
Photovoltaic energy systems.

Bibliography: p.
Includes index.
1. Photovoltaic power generation. I. Title.
TK2960.B87 1983 621.31'244 82-17150
ISBN 0-07-008952-3

1234567890 KGPKGP 89876543

ISBN 0-07-008952-3

The editors for this book were Jeremy Robinson and Susan Thomas, the designer was Elliot Epstein, and the production supervisor was Teresa F. Leaden. It was set in Baskerville by J.M. Post Graphics Corp.

Printed and bound by The Kingsport Press.

Contents

Preface

Until recently, solar cells have been little more than a popular curiosity to anyone but those specialists involved in related theoretical and laboratory work. The literature about solar cells has therefore been either esoteric or visionary. Books of all types have been published which explain the physics of light and its impact on different PN junctions, or which eloquently discuss the value of renewable resources and the important role that solar energy can play in the future of mankind. The state of photovoltaic technology has been such that little has been written to bridge the gap between hopeful ideals and practical reality. Recent progress in the field, however, is changing the picture. A combination of government and business efforts has brought the price of photovoltaic electricity down substantially and has led to the formation of small-scale industries that are pioneering the terrestrial application of solar cells and establishing a market for their wares. Concomitant with these developments is a growing demand for basic information about how solar cells can be integrated into a power system that will satisfy real energy demands. This book is based on the practical and theoretical experience developed at M.I.T. Lincoln Laboratory, Sandia Laboratory, and the Jet Propulsion Laboratory in the field of photovoltaic-system design and is intended to aid this dawning industry.

The widespread application of photovoltaic systems will require that the price be competitive with other forms of energy. The most expensive and important element in such a system is the solar cell. Most photovoltaic cells are composed of basic materials that are abundant throughout the world (commercial solar cells, to date, are generally made from silicon, which is the second most plentiful element on earth). The cost of a solar cell is largely determined by the manufacturing process, which itself requires energy. Various industries and governments are committing large amounts of money and talent to the scientific and technical innovation needed to develop a more efficient manufacturing technique. Other significant photovoltaic system expenses are for power conditioning equipment; wiring and mounting hardware; and labor costs for design and installation. The last cost factor can be reduced by simplifying and standardizing system designs so that the design, manufacturing, and installation process becomes easy to understand and implement. Ultimately, cost-effectiveness will be achieved only through mass production.

Photovoltaic electricity will make a significant contribution to a nation's energy demand when the installation of photovoltaic systems becomes as routine as the installation of electrical wiring in a building. Standard designs could be made for different regions of the country, and builders and electricians could perform both the layout and the installation work. The objective of this book is to simplify and organize the extensive information available on photovoltaic system design in a way that facilitates the standarization process.

This book focuses on the design and installation of small to intermediate-scale photovoltaic systems (100 to 100,000 peak watts) with fixed, flat-plate photovoltaic modules. Techniques involving sunlight concentration, sun tracking, and combined photovoltaic-thermal collection are not discussed here for reasons listed below. The potential applications of photovoltaic power in remote locations and on the low-cost roof space of many buildings will be great as the price of solar cells becomes increasingly competitive. Fixed flat-plate photovoltaic arrays will prove most attractive in the latter designs because of their simplicity, reliability, and low maintenance cost. Sun tracking is considered attractive because it can increase the solar energy received by as much as 50 percent. Sunlight concentration is heralded because the presently expensive cell area can be reduced. Combined photovoltaic-thermal collection at first makes good sense because most of the energy not converted into electricity goes into heat. Why not simultaneously tap this additional resource? Despite all these favorable features, the three techniques mentioned have major flaws.

Sun tracking and concentrator systems typically need a mechanical assembly to follow the sun's direct-beam rays. This assembly uses additional energy to operate and will involve considerable maintenance. Photovoltaic systems involving high concentration ratios or combined photovoltaic-thermal collection require pumps and plumbing to circulate the liquid used to store heat. Such systems can only heat the fluid significantly by forcing the solar cell to operate at a high temperature, and therefore at a low efficiency. There is a trade-off in a PV/T collector between obtaining maximum heat and maximum electric power; since both cannot be optimized simultaneously, the value of such a system decreases. Solar heating and solar electricity are best generated side by side or together in a special integral-mount passive design, discussed in Chapter 9. The fixed flat-plate photovoltaic collector does not track the sun, taps only a percentage of the heat received from the sun, and requires large amounts of cell area. It does, however, afford a simple design that can be reliable and virtually maintenance-free. Once the expected photovoltaic cell cost reductions are achieved, the value of simplicity will be critical, especially to the manufacturing, construction, and housing industries.

This book discusses the performance and design of photovoltaic systems involving *silicon* solar cells (be they single-crystal, polycrystalline, or amorphous). The photovoltaic effect can be produced using other materials, such as gallium arsenide or cadmium sulfide; to date, however, silicon is by far the most common material used in commercially available solar cells. Most of the principles presented here will be transferable to other types of PN junctions.

While this book is written at a level that is instructive to the professional engineer and architect, it is also comprehensible to the educated layperson and skilled craftsperson. The first three chapters present an introductory review of electricity demand, the photovoltaic effect, and solar energy. The major technical discussion is found in chapters four through nine and in the appendixes.

Acknowledgments

My three years of work with notable engineers involved in the research and development of standalone and utility-interactive photovoltaic systems at M.I.T. Lincoln Laboratory has formed the basis for much of the material presented here. Many of the systems developed in this research group have successfully operated for three or more years and could become models for a potentially large industry. The information presented in this book is based on the accomplishments of such pioneering scientists and engineers as Drs. Charles Cox, Steve Forman, Edward Kern, Emmanuel Landsman, Marvin Pope, Ron Ross, Al Millner, Ron Matlin, Larry Bucciarelli, John Cremin, Bud Lyon, Carl Much, P. Raghuraman, and John Solman, along with Burt Nichols, Niel Rasmussen, Tom Warner, Bonwyn Brench, Miles Russell, Susan Hendrie, Howard Branz, Joseph Doucet, Henry Bulwinkel, Michael Wong, George Hart, Rick Cadieux, Pat Phemelis, Ray Hopkinson, Harvey Fenton, Jack Helfrich, Tom Emberly, Steven Strong, and Peter Mahoney. The latter half of chapter 2 was ably written by Robert Lempert. The manuscript was carefully read and influenced by Dr. Charles Cox of M.I.T. Lincoln Laboratory and Dr. Ron Ross of the Jet Propulsion Laboratory; I am most grateful for their technical advice and editorial comments, which immeasurably improved this work. Additional reviewers were Jeffrey Cook, Richard Livingstone, and Martin Self.

I am especially thankful to Dr. Charles Cox, Victoria Cox, and my mother for their continuous moral support, and to Bob Harris for his initial ideas and encouragement. Above all, I must dedicate this work to my Lord and Author, who prepared for me the opportunity to write this book and sustained me throughout the years it took to complete the task.

Matthew Buresch

1

THE DEMAND FOR ELECTRICITY

An assessment of where and how photovoltaic electricity could make a significant contribution to our energy needs is aided by obtaining an overview of energy uses as they differ throughout society. Electricity is a modern form of energy that has varied applications in four different sectors of our economy: industrial, transportational, commercial, and residential. Up until 1981, solar-cell electricity has been used for small-scale and special applications. If they are to play an appreciable part in supplying our energy needs, photovoltaic systems must be targeted toward one or more of the major electricity demand sectors. The size and distribution of the residential electrical load makes it particularly attractive for photovoltaic applications.

ENERGY NEEDS IN AN INDUSTRIALIZED SOCIETY

Electricity was first generated and used on a significant scale in the early part of the twentieth century. Since that time, electricity use has increased at a steady rate that roughly parallels the overall increase in the U.S. demand for energy, as shown in Figure 1-1. In the coming decades, electricity needs are expected to grow more rapidly than the total energy demand, as a result of advances in technology that are resulting in widespread automation. The large energy losses that naturally occur when

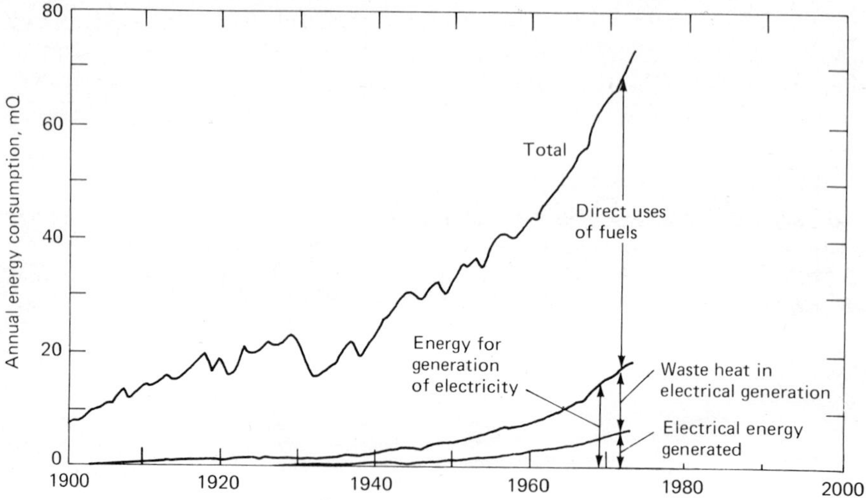

Figure 1-1 **Energy and electricity demand in the United States during the twentieth century. (From: Robert H. Romer, *Energy: An Introduction to Physics,* Freeman, San Francisco, copyright © 1976, p. 598.)**

electricity is generated in a fuel-burning plant are often outweighed by the advantages of this versatile form of energy.

Energy demands in an industrialized society can be divided into four major sectors: industrial, transportational, commercial, and residential. The relative share of the energy "pie" that each different sector requires varies according to a country's resources, technical advancement, and political and economic structure. In the United States in 1973, for example, the industrial sector required about 40% of the total national energy output. Most of this energy went into the production of chemicals, metals, and petroleum products; secondary uses included the manufacture of food, paper products, and building materials. The next largest portion (about 26%) of the total national energy supply was required for transportation. Passenger automobiles and trucks carrying freight were the dominant users of fuel in this sector. The energy demand of the residential sector (about 19%) was mostly devoted to space heating, hot water, lighting, refrigeration, cooking, and air conditioning. The smallest portion (about 15%) was used in the commercial sector for space heating and cooling, refrigeration, and hot water. The residential and commercial needs are quite similar. Table 1-1 outlines the relative energy needs of each of the four sectors in greater detail.

The energy used to supply these different economic sectors can be divided into that devoted to direct use and that converted for electric power. Table 1-2 shows how much energy was devoted to direct use and how much to electrical use in the U.S. economy in 1973. The direct use of energy involves burning fuels, such as wood, oil, gas, and coal, directly for the purpose of doing work. Direct burning of fuels is most notably used for heating space, water, or industrial materials, and for combustion in engines used for locomotion. Most electricity is created by converting energy from fossil fuels and nuclear reactions into electrical energy through the production of steam and the rotation of a generator. This conversion process is rather costly because energy is lost during generation and transmission, with over half the initial energy usually squandered in the

Table 1-1 Energy Use for Four Major Sectors of the U.S. Economy, 1973.

Sector	Direct uses of fuels,* mQ†	Electricity (including waste heat from generation),* mQ	Total energy consumption,* mQ	Percentage of national total, %
Residential	8.1 (+0.8%)	5.8 (+7.5%)	13.9 (+3.3%)	19.1
Commercial	6.7 (+3.0%)	4.5 (+7.0%)	11.2 (+4.6%)	15.4
Industrial	20.4 (+2.3%)	8.4 (+4.6%)	28.8 (+3.0%)	39.6
Transportation	18.9 (+4.4%)	0.04 (0%)	18.9 (+4.4%)	26.0
Total	54.1 (+2.9%)	18.7 (+6.0%)	72.8 (+3.7%)	100

Source: Robert H. Romer, *Energy: An Introduction to Physics,* W. H. Freeman & Co., San Francisco, © 1976, p. 603.

*Figures in parentheses are average annual percentage rates of change during the period 1970–1973.

†Q = quad = 10^{15} (or one quadrillion) Btu.

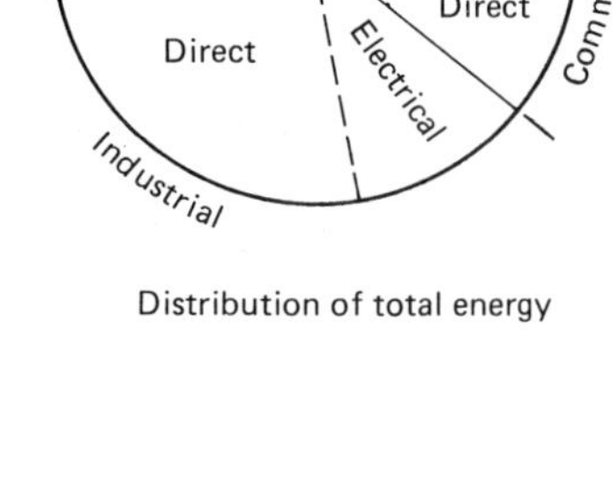

Distribution of total energy

Table 1-2 Percentages of Energy Used Directly and Converted to Electricity, United States, 1973

Uses	Energy,* mQ	Percentage of national total, %
Direct uses of fossil fuels		
Coal	4.54 (−4.5%)	6.2
Oil	31.0 (+4.5%)	42.6
Natural gas	18.6 (+2.4%)	25.5
Total direct use	54.1 (+2.9%)	74.3
Electrical generation		
Electrical energy generated	6.64 (+5.9%)	9.1
Energy wasted in electrical generation	12.06 (+6.1%)	16.5
Total for electrical generation	18.7 (+6.0%)	25.7
Total energy consumption	72.8 (+3.7%)	100

Source: Robert H. Romer, *Energy: An Introduction to Physics,* W. H. Freeman & Co., San Francisco, © 1976, p. 603.

*Figures in parentheses are average annual percentage rates of change during the period 1970–1973.

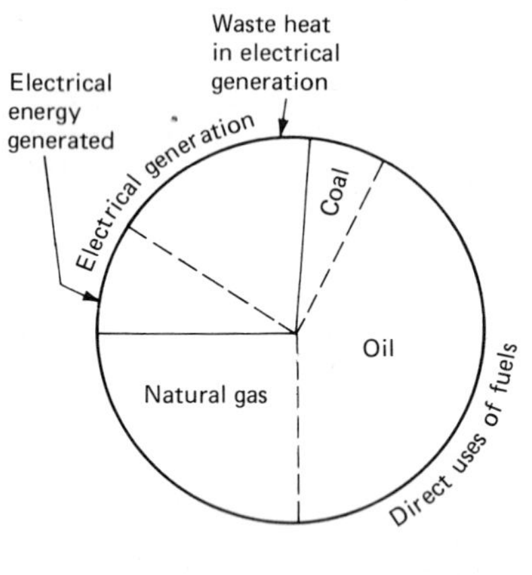

form of unused heat. Nevertheless, electricity is increasingly attractive because of its versatility for many applications that have developed since the invention of such devices as the motor, radio, and computer.

There is more use of directly burned fuel than of electricity in all four economic sectors. Transportation uses little electricity, especially in America, where electric trains and subways are not common. Most of the energy required for transportation comes directly from the burning of gasoline and diesel fuels. Only about one-third of the industrial load requires electricity, although its uses are about as varied as the overall demand. A higher percentage of energy needs are met by electricity in the commercial and residential sectors, where direct fuel and electricity are used side by side in a ratio of about 2:1.5. Direct fuel is generally used for various types of heating, and electricity is needed to operate a wide variety of appliances.

PHOTOVOLTAIC ENERGY

Where in the overall energy picture can photovoltaic electricity make a substantial impact? Clearly only those sectors that require electricity will have any use for solar cells. A look at Tables 1-3 to 1-5 and their detailed listing of direct fuel versus electrical demand in each of the three sectors provides the reader with an idea of where the use of solar cells could potentially be targeted.

Up until now, solar cells have been used for limited and very diverse applications. Satellites, electric calculators and watches, a one-person airplane, and various prototype electric cars have been designed which obtain most or all of their power from photovoltaic cells. Water-pumping irrigation projects, a water desalination plant, and a professional radio station are operating entirely on solar cells. Solar electricity has become cost-effective for remote applications where utility power is unavailable. In the southwestern United States, both an Indian village and a national park ranger station are powered by a stand-alone system of solar cells with battery storage. Radio repeater stations, railroad crossings, and mobile transmitters are operating on solar-cell power in many parts of the world.

Despite the demonstrated versatility of solar cells, there are four major limitations that could prevent them from making a significant contribution to the national energy demand. These issues must be addressed if photovoltaic generators are to be transformed from a special-application power source into a widespread alternative. The cost of solar cells, the amount of annual sunshine at a site, the type and timing of the PV output power, and the space required to generate a significant amount of photovoltaic electricity are the four most important considerations involved in the application of this form of electric power.

The cost of solar cells has decreased from about \$300 to about \$7 (1980 dollars) per peak watt (Wp) between the years 1965 and 1980, and probably will have to drop to about \$0.50 or \$1.00 per peak watt

Table 1-3 Energy Use in the U.S. Industrial Sector, 1973

Use	Direct uses of fuels,* mQ	Electricity (including waste heat from generation),* mQ	Total energy consumption,* mQ	Percentage of industrial sector, %	Percentage of national total,%
Chemicals (production of basic chemicals, synthetic fibers, drugs, fertilizers, etc.)	3.6 (+2.7%)	2.4 (+5.0%)	6.0 (+3.6%)	20.8	8.2
Metals (smelting and refining, manufacture of basic metal products)	3.9 (+0.5%)	2.0 (+4.7%)	5.9 (+1.9%)	20.5	8.1
Petroleum refining, manufacture of paving and roofing materials, lubricants, etc.	3.0 (+4.0%)	0.3 (+5.1%)	3.3 (+4.1%)	11.5	4.5
Food manufacturing and processing	1.0 (+2.4%)	0.5 (+4.0%)	1.5 (+2.9%)	5.2	2.1
Paper products	1.0 (+1.4%)	0.5 (+4.8%)	1.5 (+2.5%)	5.2	2.1
Glass, concrete, asbestos, etc.	1.0 (+1.7%)	0.4 (+4.1%)	1.4 (+2.4%)	4.9	1.9
Other industries	6.9 (+2.6%)	2.3 (+4.2%)	9.2 (+3.0%)	31.9	12.6
Total	20.4 (+2.3%)	8.4 (+4.6%)	28.8 (+3.0%)	100	39.6

Source: Robert H. Romer, *Energy: An Introduction to Physics,* W. H. Freeman & Co., San Francisco, © 1976, p. 605.

*Figures in parentheses are average annual percentage rates of change during the period 1970–1973.

Table 1-4 Energy Use in the U.S. Commercial Sector, 1973

Use	Direct uses of fuels,* mQ	Electricity (including waste heat from generation),* mQ	Total energy consumption,* mQ	Percentage of commercial sector, %	Percentage of national total, %
Space heating	4.4 (+1.2%)	1.0 (+30%)	5.4 (+4.7%)	48.2	7.4
Nonenergy uses†	1.6 (+9.2%)	0	1.6 (+9.2%)	14.3	2.2
Air conditioning	0.15 (+8.1%)	1.47 (+7.3%)	1.6 (+7.3%)	14.3	2.2
Refrigeration	0	0.68 (0%)	0.68 (0%)	6.1	0.9
Hot water	0.43 (+1.0%)	0.22 (+1.0%)	0.65 (+1.0%)	5.8	0.9
Lights (except street and highway)	0	0.47 (+0.4%)	0.47 (+0.4%)	4.2	0.6
Cooking	0.13 (+1.4%)	0.08 (+4.0%)	0.21 (+2.5%)	1.9	0.3
Street and highway lights	0	0.10 (+0.3%)	0.10 (+0.3%)	0.9	0.14
Other	0	0.48 (+0.4%)	0.48 (+0.4%)	4.3	0.7
Total	6.7 (+3.0%)	4.5 (+7.0%)	11.2 (+4.6%)	100	15.4

Source: Robert H. Romer, *Energy: An Introduction to Physics,* Freeman, San Francisco, © 1976, pp. 604–605.

*Figures in parentheses are average annual percentage rates of change during the period 1970–1973.

†Largely use of petroleum products for highway paving, etc.

Table 1-5 Energy Use in the U.S. Residential Sector, 1973

Use	Direct uses of fuels,* mQ	Electricity (including waste heat from generation),* mQ	Total energy consumption,* mQ	Percentage of residential sector, %	Percentage of national total, %
Space heating	6.4 (+0.7%)	1.26 (+24.7%)	7.7 (+3.5%)	55.4	10.6
Hot water	1.2 (+1.3%)	0.62 (+2.5%)	1.82 (+1.6%)	13.1	2.5
Lights	0	1.09 (+0.3%)	1.09 (+0.3%)	7.8	1.5
Refrigeration	0	0.76 (+4.8%)	0.76 (+4.8%)	5.5	1.0
Cooking	0.36 (0%)	0.28 (+3.6%)	0.64 (+1.6%)	4.6	0.9
Air conditioning	0	0.56 (+8.3%)	0.56 (+8.3%)	4.0	0.8
Television	0	0.42 (+5.9%)	0.42 (+5.9%)	3.0	0.6
Clothes drying	0.08 (+2.6%)	0.19 (+8.8%)	0.27 (+6.8%)	1.9	0.4
Freezers	0	0.29 (+8.3%)	0.29 (+8.3%)	2.1	0.4
Misc. appliances	0	0.22 (+2.5%)	0.22 (+2.5%)	1.6	0.3

Dish washers	0	0.06 (+10.6%)	0.06 (+10.6%)	0.4	0.08
Washing machines	0	0.04 (0%)	0.04 (0%)	0.3	0.05
Total	8.1 (+0.8%)	5.8 (+7.5%)	13.9 (+3.3%)	100	19.1

Source: Robert H. Romer, *Energy: An Introduction to Physics,* W. H. Freeman & Co., San Francisco, © 1976, pp. 604–605.

*Figures in parentheses are average annual percentage rates of change during the period 1970–1973.

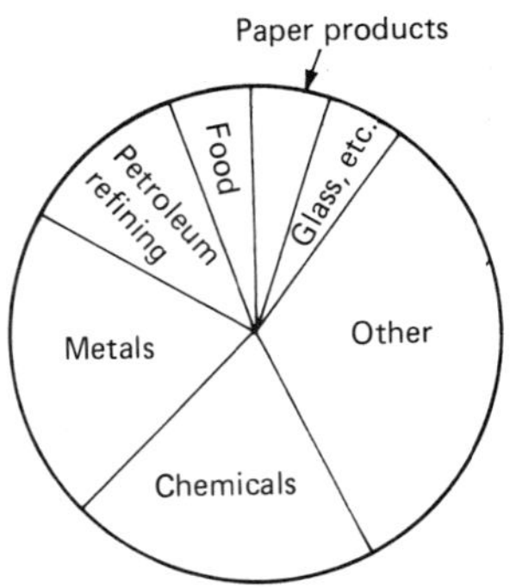

Industrial Sector

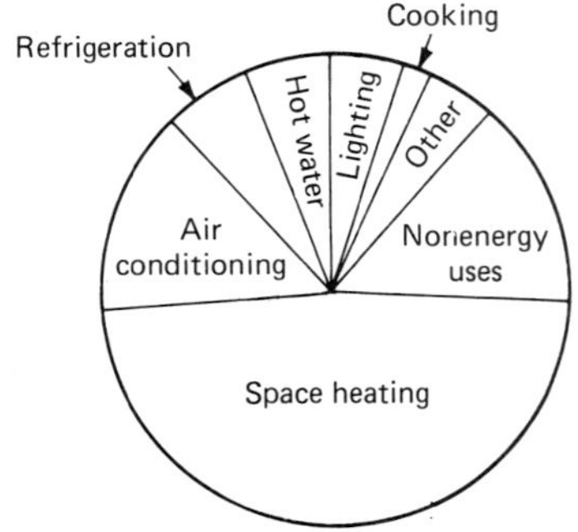

Commercial Sector

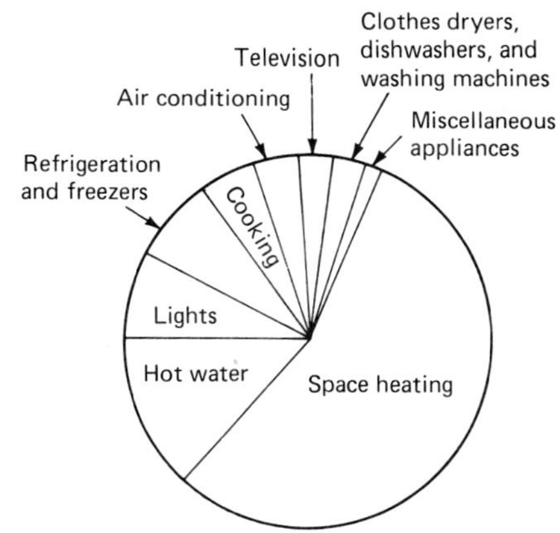

Residential Sector

before it can directly compete with utility power rates. The revolutionizing of the solar-cell manufacturing process leads experts to have strong expectations that this cost reduction will be achieved. (The technicalities of reducing the manufacturing costs are discussed in Chapter 2 and are covered in greater detail in various books listed in the bibliography.)

Additional cost factors affecting the feasibility of solar cells are the large variations in utility electricity prices in different parts of the country and the tax structure that enables certain expenses to be written off. Electricity rates can vary by as much as a factor of 6. For instance, as of late 1979, electricity in Seattle, Washington (mostly generated from hydropower), cost only $0.022 per kilowatthour. At about the same time, in New York City (which relies on oil and coal), the cost was as high as $0.139 per kilowatthour. The variety of tax credits and deductions available for the purchase of solar electric generators varies with the political climate and with the category the purchaser fits in. Solar cells clearly will be most attractive in those parts of the country where utility electricity sells for the highest price and the tax structure is most favorable.

The annual solar energy can vary by almost as much as a factor of 2 in different parts of the continental United States. For instance, in Phoenix the typical annual solar energy received on a horizontal surface is about 2330 kWh/m^2, while in Boston the solar energy obtained over the same period is about 1380 kWh/m^2. Chapter 3 will demonstrate how solar energy can be predicted for different sites. Clearly, those places that experience the most solar energy will be the most attractive for establishing a photovoltaic electricity-generating facility.

The timing and form of photovoltaic power can pose a problem. Solar electricity is generated only during sunny daytime hours, while most applications need a dependable supply of electricity that is also available during cloudy periods and at night. The voltage output of a photovoltaic array may fluctuate to intolerable levels, and the direct current that is intrinsically generated by a solar cell often is incompatible with the more conventional loads powered by alternating current. The resolution of these issues could involve energy storage, voltage regulation, and DC-to-AC conversion, as will be discussed in Chapter 7. Those systems that can minimize these power-conditioning requirements will be the most cost-effective.

The space required to generate a significant amount of photovoltaic electricity at first seems to be another problem that might restrict the use of solar cells. Photovoltaics require area with continuous exposure to the sun at the rate of one square meter for every one hundred to two hundred watts of peak output power. By comparison, most conventional fossil-fuel or nuclear electric power plants require one square meter of land area to generate about six thousand to eight thousand watts. A solar cell provides peak power only in the middle of a sunny day, while fuel-burning power plants sustain their power level at all hours. An array of solar cells capable of producing as much peak electric power as a conventional electricity-generating facility would require between 5 and 15 km^2 of uninterrupted cell area. Yet in order to keep transmission losses to a minimum, it is desirable to situate power plants as close to their load as possible. Since most of the nation's electric loads are located in urban areas where the price of real estate is high, space constraints make the use of solar cells for large central power-generation facilities less attractive. There have been suggestions that solar-cell power plants be put in orbit and electricity beamed back to earth via microwaves, but this controversial undertaking would require a huge initial investment.

An inspiring solution to the space problem may be found by looking at a city from a bird's-eye (or, more appropriately, a sun's eye) view. There is a large amount of area in a city that is continuously exposed to the sun, yet upon which people have not placed a price tag: the roof area of any house of building. Although some sloping roofs do not face south and others are shaded most of the time by taller buildings or trees, the adequately positioned roof area is substantial in suburban locations and parts of cities without high buildings. The availability of this area suggests a solution to the space problem; photovoltaic electricity is attractive for residential and commercial applications.

THE RESIDENTIAL ELECTRIC LOAD

Our study of photovoltaic applicability will concentrate on the residential electric load for two reasons. First, a one-family residential electric load can be entirely or in large part satisfied by the photovoltaic electricity-generating capability of a residential-sized roof. Second, an introductory study of electric loads is best performed on devices that are most familiar, namely household appliances.

The size and distribution of residential loads make them well suited for the application of solar cells. With the exception of large apartment houses, individual residences occupy a sizable amount of land area and have considerable roof space exposed to the sun; the load requirements for a home are small compared with the area that a house occupies. In fact, if a south-facing roof of a typical one-family house were covered with solar cells, the photovoltaic array could provide a large portion, if not all, of that residence's electrical needs. The electric energy demand of a one-family residence without electric space heating or cooling typically is between 5 and 35 kWh/day; the winter energy use is on the high end of the scale (days are shorter, so that more lights are on and more time is spent indoors), while the summer is on the low end. The non-heating or cooling residential appliance load does not vary substantially among different parts of the United States. If a 50-m^2 roof area with a south-facing tilt were covered with photovoltaic cells, it could generate between 12 and 18 kWh of electricity per average annual day in Boston and between 20 and 30 kWh in Phoenix.

The residential load includes a variety of appliances that are very familiar. There is hardly a person in the industrialized world who has not regularly spent time using a refrigerator, television, radio, or washing machine. In the near term, the major use for photovoltaic energy is for powering remote-site industrial and commercial loads. Knowledge about the residential electrical load, however, will be directly transferable to these electric-power applications. For instance, the electricity requirements of a refrigerator in a home and a radio repeater station on a mountain peak are assessed in the same manner.

The electricity demand of a residence is neatly summarized each month in the electric utility bill, which indicates the amount and cost of the power used during the billing period. A curious homeowner might compare the energy needs for different months of the year or for different homes in the neighborhood. In most cases, however, the total dollar figure on the bill is given more attention than the measured kilowatthour reading necessary to make a cost calculation. Neglect of such energy information is partially due to the difficulty of taking a lump-sum energy figure for a particular month and breaking it down into the various energy and cost requirements of each home appliance. It is conceptually difficult to connect a number on a bill with the many switches and devices we use each day.

A good technique for analyzing the electrical demand of a single residence load is presented in the following five-step program. Each residence has its own arrangement of electric loads, and each inhabitant has a unique lifestyle. In order to summarize the electric energy needs of a household, the amount and time of use for each device or load must be considered.

The first step is to make a comprehensive list of every electric-powered load used in the home. A tour through each room, tracing all appliances connected to wall outlets or wall-mounted switches, can make this listing of individual loads a systematic and simple task. Especially important items are boilers, freezers, refrigerators, space heaters, clothes dryers, and cooking ranges; if powered with electricity, these will constitute a major part of the residential electricity demand.

The second step is to determine the power ratings for each appliance listed in step 1. The power rating is measured in watts (W) and is usually printed or labeled on the back of each device. Finding these markings can involve moving large appliances such as refrigerators and dishwashers out of their normal place, climbing up to lighting fixtures to examine light bulbs, wandering into the basement to check washing machines or electric boilers, and exploring the outside of a house for lights and electric outlets. Many homeowners do not have sufficient motivation or physical resources for such an adventure. An alternative is to find the power ratings for appliances by examining owners' manuals or by looking up typical figures in a table of power ratings, such as Table 1-6.

The third step is to estimate the *ON time* and note the time of day when appliances are in use. The electricity requirements of some appliances vary during different times of the year, and therefore it may be necessary to repeat this exercise for a typical day in each of the four seasons.

Very few devices are actually on continuously for an entire day (one common exception is the electric clock). Most appliances are manually or automatically switched on and off at various times during the day and night. The ON time for such manually operated devices as lights, televisions, radios, washing machines, and vacuum cleaners is easier to estimate because an individual is aware and in control when the appliance is used. Automatically controlled machines such as refrigerators, circulating fans, air conditioners, and electric heating systems are turned on and off by temperature controllers called *thermostats,* which are designed to maintain a desired temperature level. (For example, a refrigerator or air-conditioner thermostat turns on its compressor and pumps Freon through the cooling chamber when the temperature rises above a desired level, turning the compressor off when the chamber reaches the controlled level of coolness. Thermostats for electric heating elements work in the opposite direction, turning off the heating coil when the temperature exceeds a desired level and turning it on when the temperature drops too low.) Thermostats are usually adjustable, so that various settings can be chosen. The automatic ON time thus depends on the temperature setting, the ambient temperature, the space to be cooled or heated, and the appliance's efficiency; and it can fluctuate from one device or situation to the next. Semiautomatic appliances such as coffeemakers, electric ovens, and clothes dryers are started manually but thereafter are controlled by thermostats which cycle them off and on.

Take the list established in step 1 and write in the estimated ON time for each electric appliance during a typical 24-hour period. Note also

Table 1-6 Power and Energy Requirement for Electric Appliances

Appliance	Average power required, w	Estimated electric energy used per year, kWh	Appliance	Average power required, w	Estimated electric energy used per year, kWh
Air conditioner (window, 5000 Bu/h)	1565	1390	Humidifier	175	163
Blanket	177	147	Iron	1000	144
Broiler	1436	100	Microwave oven	1450	190
Carving knife	92	8	Radio	71	86
Clock	2	17	Radio-phonograph	110	110
Clothes dryer	4855	990	Razor	14	2
Coffee maker	895	105	Refrigerator (12 ft^3)	240	730
Deep-fat fryer	1450	83	Refrigerator (12 ft^3)	320	1215
Dishwasher	1200	363	Refrigerator-freezer (14 ft^3)	325	1135
Egg cooker	516	14	Refrigerator-freezer (14 ft^3 frostless)	615	1830
Fan (attic)	370	290			

Fan (circulating)	88	43	Roaster	1333	205
Fan (window)	200	170	Sewing machine	75	11
Floor polisher	305	15	Stove	12,200	1175
Food blender	385	15	Sun lamp	280	16
Food freezer (15 ft^3)	340	1200	Television		
Food freezer (15 ft^3 frostless)	440	1760	B&W, vacuum-tube	160	350
			B&W, solid-state	55	120
Food mixer	127	13	Color, vacuum-tube	300	660
Food waste disposer	445	30	Color, solid-state	200	440
Frying pan	1200	185	Toaster	1145	40
Grill (sandwich)	1160	33	Toothbrush	7	0.5
Hair dryer	380	14	Trash compactor	400	50
Heat lamp	250	13	Vacuum cleaner	630	46
Heater (portable)	1320	175	Vibrator	40	2
Heating pad	65	10	Waffle iron	1115	22
Hot plate	1260	90	Washing machine	512	103

Source: Electrical Energy Association, New York; in Robert H. Romer, *Energy: An Introduction to Physics,* W. H. Freeman & Co., San Francisco, © 1976, p. 608.

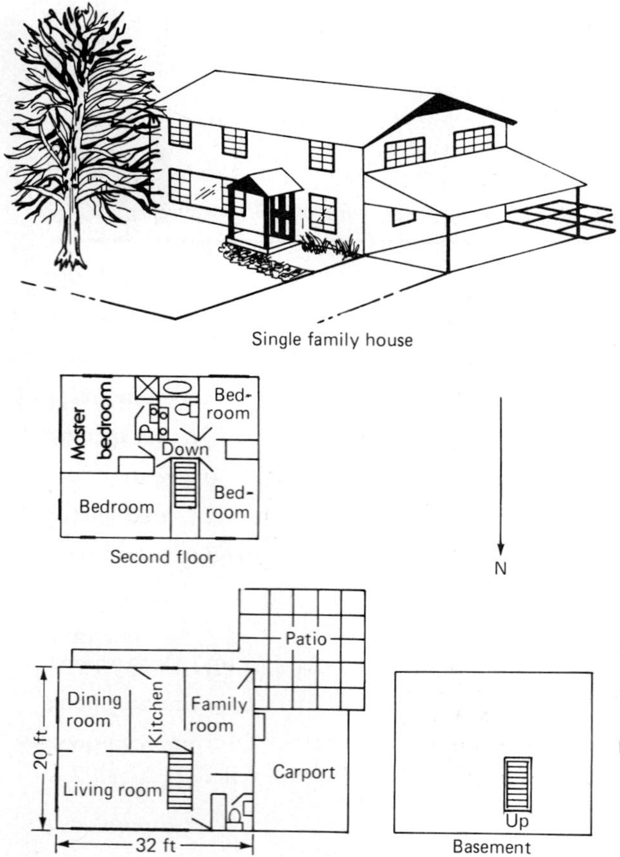

(a)

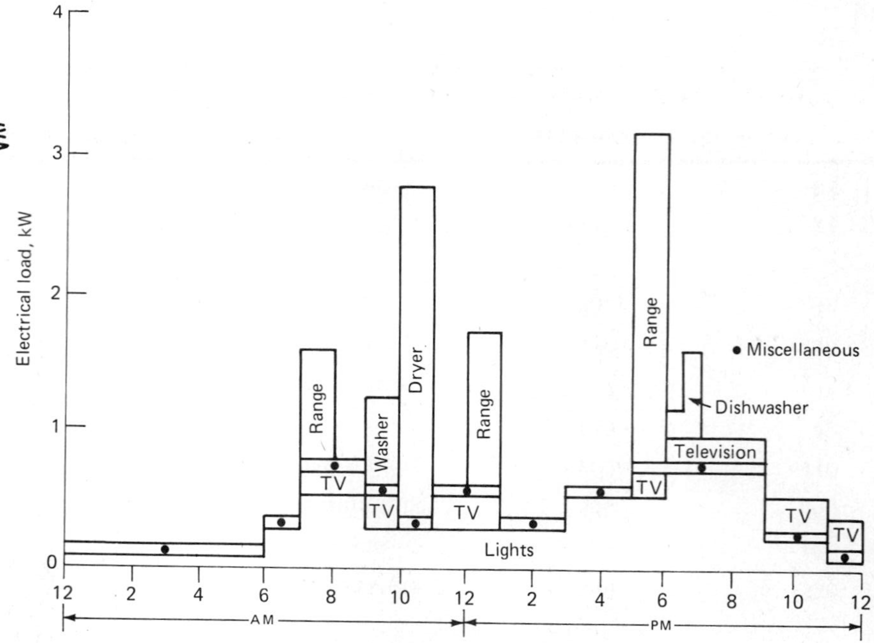

Note: Short peaks may exceed these hourly averages.

(b)

Single-Family Occupancy and Demand Profiles

	Occupancy profile	Appliance & lighting electricity, kWhe	Annual average hot water use kWh_t		Occupancy profile	Appliance & lighting electricity, kWhe	Annual average hot water use kWh_t
1	4.0	0.15	0.0	15	1.6	0.35	0.0
2	4.0	0.15	0.0	16	1.6	0.55	0.0
3	4.0	0.15	0.0	17	1.6	0.55	0.0
4	4.0	0.15	0.0	18	2.8	3.25	1.0
5	4.0	0.15	0.0	19	2.8	1.35	2.51
6	4.0	0.15	0.0	20	3.6	0.95	2.51
7	4.0	0.35	0.5	21	3.6	0.95	1.0
8	3.6	1.55	2.01	22	3.6	0.50	0.0
9	2.8	0.75	3.01	23	4.0	0.50	0.0
10	2.8	1.25	5.02	24	4.0	0.35	0.0
11	1.6	2.85	5.02				
12	1.6	0.55	0.0	Daily total		19.60	23.07
13	1.6	1.75	0.5				(105 gal/d)
14	1.6	0.35	0.0	Yearly total		7.154	8.422

Figure 1-2 Electric power use versus time of day for a typical single-family house without electric heating and cooling. (*a*) Plan of characteristic house. (*b*) Assumed appliance and lighting electrical-load profile. (Office of Technology Assessment, *Application of Solar Technology to Today's Energy Needs*, Congress of the United States, pp. 701, 702.)

the time of day when each device is used. ON time for manually controlled appliances can be easily estimated by observing the daily habits of people in the house. The ON time for semiautomatic appliances can be roughly determined by observing how long the device was on manually and then estimating the automatic ON time by noting the clicking sound, pilot-light flickering, or light dimming associated with the on/off cycling of the thermostat. The ON time for fully automatic appliances such as refrigerators or central heating systems may be more difficult to ascertain, but it can be estimated by watching or listening to the appliances for signs of thermostat cycling. An alternative is questioning the manufacturer or a sales and repair shop. Some appliances are used only a few days out of the month; their time in use must be averaged out to provide a daily time figure.

A graphical representation of the power versus time of day provides a good picture of how the residential load varies throughout the day. The graph and information provided in Figure 1-2 portray the power/time-of-day plot for a typical single-family house. Notice how certain devices are primarily used during well-defined periods of the day, while others are on almost continuously.

The fourth step in estimating the electric-energy demand involves computing the energy used by each appliance. The distinction between power and energy becomes clearer at this point. Since power represents energy transacted per unit time, multiplying power used by a unit of time will equal total energy.

$$\text{Power} \times \text{time} = \text{energy/time} \times \text{time} = \text{energy}$$

To compute the energy used by a household appliance (in kilowatt-hours), multiply its power rating (in watts) by the total ON time (in hours) during a 24-hour period. The result will equal each load's daily energy use in watthours. Note that some appliances with high power ratings and short times in use (such as toasters) may use less energy than televisions or other loads which have fairly low power ratings but are used often.

The total electric-energy demand for one day is the sum of all the energy used by all the appliances. To compare this energy use estimate with what the utility has measured, divide the utility's monthly energy figure in watthours by the number of days in the billing period. There will probably be a discrepancy between the home-calculated figure and the utility's measured figure for a typical day's energy use, but this difference should be small if the estimations and calculations were performed with care.

The fifth and final step in our energy assessment is to determine the monthly average daily energy demand for each month of the year. The electricity used by a residence varies throughout the year because of seasonal changes in temperature and in the number of daylight hours. Taking the kilowatthour energy reading on each month's bill and dividing by the exact number of days in the billing period will yield the monthly average daily electricity demand. Plot these monthly average daily energy figures against the month to produce a simple graphic display of how the residence's energy needs vary throughout the year.

The energy/time-of-year curves for residences in different locations and climates can be quite dissimilar; houses that are equipped with dif-

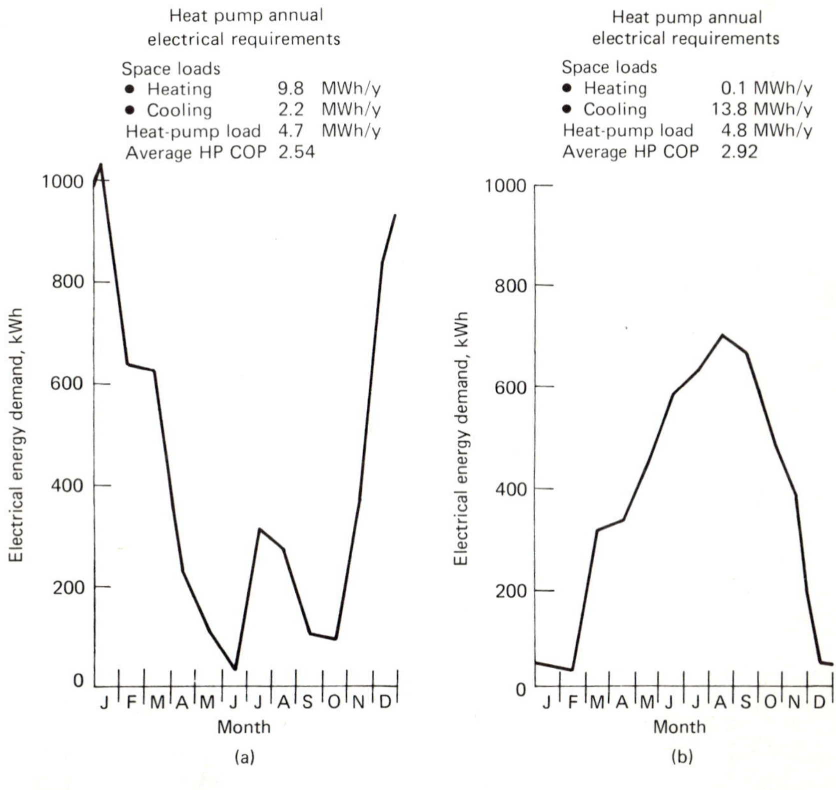

Figure 1-3 Energy-demand/time-of-year curves for electric heating and cooling with a heat pump, (*a*) for a Boston and (*b*) for a Miami single-family residence. (E. M. Mehalick, et al., *The Design of a Photovoltaic System for a Southwest All-Electric Residence*, Sandia Laboratory, Albuquerque, N. Mex., 1980, p. 6–4.)

Table 1-7 Sample Northern-Climate Residence Electric-Energy Use on a Typical Spring Day.

1	2	3												4	5
		Time in use, h													
Appliance	**Wattage rating, W**	**12–2 AM**	**2–4 AM**	**4–6 AM**	**6–8 AM**	**8–10 AM**	**10–12 AM**	**12–2 PM**	**2–4 PM**	**4–6 PM**	**6–8 PM**	**8–10 PM**	**10–12 PM**	**Total time in use**	**Energy, Wh**
Stove	12,200*				0.1							0.2		0.3	3,660
Clothes dryer	4,855									0.05				0.05	243
Toaster	1,145				0.1									0.1	114
Washing machine	512								0.05					0.05	25
Hair dryer	380											0.25		0.25	95
Lights	1,260										0.5	1.5		2	2,580
Frostless refrigerator	320	0.7	0.7	0.7	0.8	0.7	0.7	0.7	0.7	0.7	0.7	0.8	0.7	8.9	2,848
Vacuum cleaner	630						0.1							0.1	63
TV, color	200				1						1	2		4	800
(2) Stereo radios	142*				1							1	1	3	426
(3) Electric clocks	6	2	2	2	2	2	2	2	2	2	2	2	2	24	144
Total daily energy, Wh															10,938

**Note:* These ratings assume that the device is on full blast, which generally is not the case. Adjust the time in use accordingly.

ferent quantities of electric appliances or whose inhabitants have different lifestyles will also differ in their energy demand profiles. If any type of space heating or cooling is performed with electricity, the residence's geographic location will play a large part in determining how much energy is needed and when. In the southern part of the United States, many residences require a maximum amount of electricity during the summer months to run air conditioners for cooling, while in the North the peak electricity usage typically occurs in the winter for the purpose of heating. Figure 1-3 shows the contrasting residential-energy-demand/time-of-year curves for a northern and a southern residence.

A sample problem involving a one-family house in the northern United States during the springtime is given in Table 1-7 and Figures 1-4 and 1-5. The first step involves listing all the appliances, as shown in column 1 of Table 1-7. The second step entails determining the power ratings of each and every appliance, as shown in column 2 of the table. The third step is to list the time of day and total time in use for each appliance, as performed in columns 3 and 4. A graph of residential load power versus time of day can then be plotted, as shown in Figure 1-4. The fourth step involves computing the daily energy required by each ap-

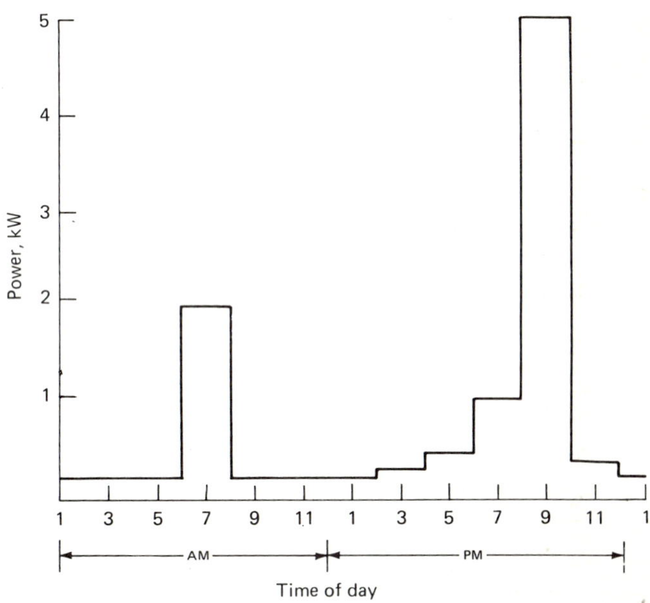

Figure 1-4 Electric-power use versus time of day for a sample residence on a typical spring day in a northern climate.

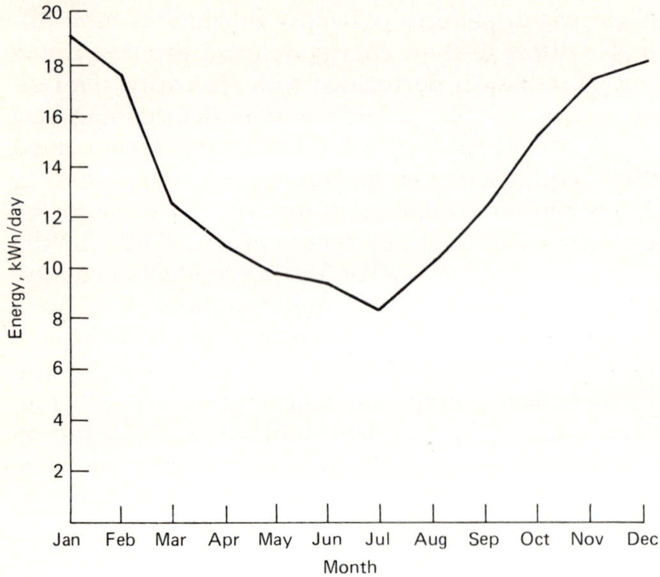

	Input Monthly Average Daily Energy, kWh/day
Jan	19.10
Feb	17.60
Mar	12.50
Apr	10.90
May	9.80
Jun	9.30
Jul	8.30
Aug	10.10
Sep	12.30
Oct	15.20
Nov	17.40
Dec	18.10
Avg. daily energy	13.38

Figure 1-5 **Electric energy versus time of year for a sample residence in a northern climate.**

pliance and the entire residence, as shown in column 5. To do this, multiply together the figures in columns 2 and 4 (power × time = energy). Finally, in step 5, determine the residential monthly average daily electricity demand for each month of the year and then plot this information on a graph, as shown in Figure 1-5.

2

GENERATING ELECTRICITY FROM SUNLIGHT*

Solar cells convert the sun's light directly into electricity; heat plays no constructive part in the process. Heat actually hampers the performance of these thin wafers, and the presence of excessive warmth is an indication of their failure. Most existing ways of generating electricity—whether oil, coal, or nuclear—involve a heating cycle in which steam is generated to drive a turbine. Even a hydroelectric plant relies on the sun's heat to transport water from lake and ocean to the highlands. The photovoltaic process of generating electricity is puzzling in that it absorbs something from light that is not heat. Understanding the solar-cell technique involves understanding how the sun's energy is produced, how it reaches the earth, and what kind of impact it has on substances when it reaches the earth's surface.

In early times the sun was believed to be a hot ball of fire, but when scientists accurately began to measure the amount of energy released by the sun as compared with its volume, they quickly realized that the sun would have long since burned up if its power were due only to a chemical reaction such as combustion. Chemical reactions involve the rearrangement of atoms into various bonded structures called molecules. For instance, when most substances burn, the carbon atoms in the flammable substance combine with the oxygen atoms in the air to form molecules of carbon dioxide (CO_2) and a residue of ash.

Flames are the evidence of atoms bonding with such rapidity that energy stored in the flammable matter is released in the form of heat

*The latter part of this chapter, beginning on page 35, was written by Robert Lempert.

and light as many atoms venture into a lower energy state. The light and heat produced during combustion may seem great, but imagine how insignificant they would be if they were 93 million miles away. If the sun were the size of a plump grapefruit, the earth would be no larger than a shriveled pea orbiting at a distance of 25 yards. Set that grapefruit on fire and the pea would hardly notice it. Another explanation had to be sought for the sun's vast energy output, and during the early part of the twentieth century, exploration into the structure of matter began to provide some clues.

THE ATOMIC STRUCTURE OF MATTER

In 1913 the Danish physicist Niels Bohr conceived a simple model of an atom as a small-scale solar system in which electrons orbit the nucleus much as the planets revolve around the sun. Subsequent research has modified Bohr's conception of the atom. Electrons spinning around a nucleus are maintained in their orbits by the mutual attraction of opposite electric charges. The negatively charged electron is prevented from randomly darting off into space by the attractive force of the positively charged proton. The proton is larger and more massive, and hence it becomes the nucleus or center about which the electrons revolve. On the other hand, the planets orbiting the sun are held in their elliptical path by the force of gravitation. Unlike electric charge, gravitation has no polarity and exists between all substances that have any mass. Since the planets have mass, they have a small attraction for one another. Electrons, being all of negative charge, repel one another. Nevertheless, drawing parallels between the earth and an electron, the sun and a proton, and their respective orbits about a center helps to visualize the physical structure of the atom.

The electrons are arranged in structured shells about their nucleus. Each shell can hold only a certain number of electrons, and when it has them it sends any newcomers to the next outer shell. Atoms have an irresistible urge to fill their outermost shell of electrons. For instance, a small atom which has seven electrons in an outer shell that can hold eight has a great attraction for any free or available electrons. Such an atom is aptly called an *electron acceptor.* On the other hand, an atom which has only one electron in its outer shell has a poor hold on this electron, and as a result is quite likely to lose it. Such an atom would be called an *electron donor.* Metal atoms are electron donors and therefore are good conductors of electricity. Nonmetallic atoms such as chlorine are electron acceptors and resist the flow of electrons, thus making them good insulators. The proton is about 1800 times more massive than the electron and hence is a less active member of the subatomic community. The electron is in perpetual motion and performs the major part in establishing bonds with other atoms. When atoms bond together in molecular formations, they generally share their outermost or valence electrons in such a fashion as to completely fill the outer shell of each atom.

Rampant consumers of energy among us might be relieved to learn that energy is neither created nor destroyed. The universe is custodian of a fixed amount of it. The impression we so often get of energy being generated or spent is in fact no more than the transfer of energy from a concentrated or useful state to a diffuse or less beneficial form. Energy can also be transformed into or extracted from matter. These transfers of energy can be explained by events occurring within the atom.

The different shells inhabited by electrons dancing about the nucleus of an atom represent energy levels. When an atom is exposed to sufficient energy in the form of light or heat, electrons can be elevated to a shell further away from the nucleus. Contrariwise, when an electron falls back into a lower orbit, energy in the form of light or heat is released, as shown in Figure 2-1. The energy absorbed or released exists in discrete bundles called *quanta*. The quantum of energy required to elevate an electron of a particular atom from the first electron shell to the second is identical to the quantum released if an electron falls back from the second to the first shell.

Imagine if you will that an electron is a salmon and the shells are small waterfalls in a river. As the salmon battles its way upstream to reach its original spawning grounds, it must surmount these various waterfalls. In order for our salmon to clear one of these falls, it will have to exert itself to make a sufficiently large flip of its tail. If its tail motion is not energetic enough, it will be snatched by the torrents of the waterfall and swept back down to its starting point. We could categorize our salmon tail-flip energies as discrete quanta in the sense that it takes a given tail flip for the struggling fish to ascend a particular waterfall.

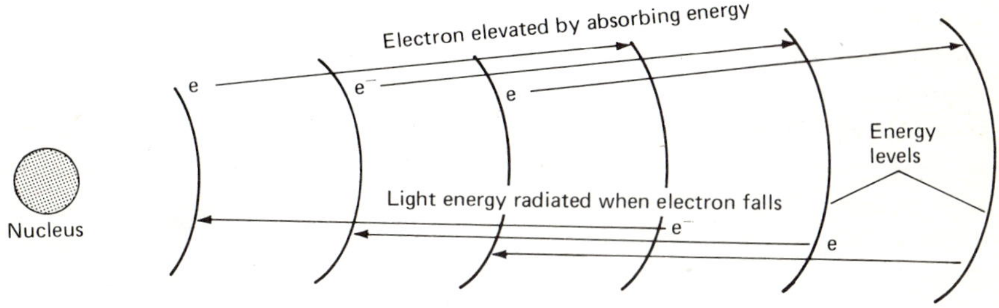

Figure 2-1 **Electron shells about an atom's nucleus represent energy levels.**

Returning from river and fish to orbit and electron, we should note that it takes smaller quanta to raise an electron from the second shell to the third than it does to raise it from the first to the second; the farther we get from the nucleus, the smaller is its influence. The outermost or valence electrons exist in what is called the *valence band.* The area beyond the valence band is called the *conduction band,* and it is in this region that liberated electrons wander at will. The energy barrier between the valence band and the conduction band is called the *band gap;* this gap is the "waterfall" an electron must surmount in order to become liberated. The quantum of energy needed to lift an electron across the band gap varies from one atom to the next, for as we move up the Table of Elements to larger atoms we find the number of electrons and their shells increasing. As the electron population grows, they interact somewhat like people in a crowd: they nudge and elbow one another to establish their place. These repulsive forces between electrons add to and subtract from the quantum figure in such a way as to make it different for each atom. An atom has its own "fingerprint" or identification based on the quanta associated with moving electrons in its shells.

RADIATION AND THE SUN

Energy is released from or absorbed by atomic structures in the form of electromagnetic radiation. This radiation has the properties of both waves and particles.

The wave characteristics are best exhibited by sunlight passing through a prism and displaying a rainbow spectrum of its component rays. Viewing it as a wave, we notice that electromagnetic radiation varies in the number of cycles per second, or frequency. Since we know the speed of light to be 186,000 miles per second, we can calculate the wavelength of any particular frequency of radiation. This radiation has different properties as its wavelength changes from that of light waves to that of radio waves, to x-rays, to radar waves, to microwaves. The only factor that separates the light waves illuminating our world and the radiation upon which our radio and television signals reach us is the wavelength.

One particle-like property of electromagnetic radiation is that it interacts with matter in quanta called *photons.* When we speak of light striking a solar cell, we more generally refer to photons with different amounts of energy hitting its atoms. If a photon has sufficient energy, it can cause an electron to leap up to a shell further from the nucleus, or even to become a free electron.

Trying to integrate the wave and particle properties of electromagnetic radiation, we could say that the waves indicate a probability distribution of photons at any instant in time. At the crest of a wave, we would find a far greater quantity of photons than at the trough.

If the sun's energy is not generated by a chemical reaction, then what might its source be? Near the end of the nineteenth century, physicists began examining the interior makeup of the atom. It was discovered that the energy holding the nucleus of an atom together was of the order of a million times greater than the energy holding electrons in their orbit. If an atom's nucleus could be split or fused, we might release a phenomenal amount of energy compared with what is produced when electrons are maneuvered about. Altering an atom's nucleus in effect transmutes one element into another and produces a nuclear reaction. The sun is a massive nuclear reaction, producing temperatures that range from 600 to 1,000,000°C. Atoms in such a cauldron are smashed about so violently that they do not remain intact. In parts of the sun the temperature reaches such a point that most of the electrons are ripped off an atom, and protons and electrons swirl about in a nondescript plasma. In such a dynamic environment, atoms are easily transmuted from one element into another as protons combine or separate. It is generally believed that the sun fuses hydrogen atoms together to form helium atoms. Hydrogen has one proton in its nucleus, while helium has two protons. The most common form of helium also has two neutrons in its nucleus (a neutron is a subatomic particle of neutral charge that under certain conditions is capable of disintegrating into a proton and an electron). Consequently, it would take four hydrogen atoms to form one helium atom.

The fusion reaction releases vast quantities of energy in the form of electromagnetic radiation. This radiation travels 93 million miles through vacuous space, is filtered by the earth's atmosphere, and finally strikes our home. The visible portion of this radiation, known as *light,* plummets toward matter in discrete particles called photons; the most easily perceived effect of this bombardment is a rise in temperature. The energy from light increases atomic and molecular vibration; we experience this random jittering of atoms as heat. More notably, photons also excite electrons into higher energy levels, thus producing free electrons. The task of a solar cell is to give these photogenerated free electrons an impetus in a productive direction.

THE PN JUNCTION: A BUILT-IN ELECTROMOTIVE FORCE

Let us introduce the main components in our scheme to generate electricity from sunlight. The major role is performed by the element silicon, followed by two antagonists called phosphorus and boron.

Silicon is the second most common element on the face of the earth and has a tendency to form stable crystalline molecules such as those commonly found in rocks and beach sand. Most solar cells are made of a purified silicon; the presence of unintentional impurities inhibits their performance. As of now, the most common photovoltaic cells are composed of a thin wafer of single-crystal silicon (about 0.1 cm thick) that is brittle like glass and cracks easily when stressed. The active electricity-generating region of a solar cell is only about 0.0001 cm thick; the remaining silicon provides structure.

The best solar cells are made from one carefully grown crystal in which all the atoms are facing the same direction and fit together into a neat three-dimensional grid. Multicrystal or polycrystalline cells are less expensive to grow but have a lower conversion efficiency. Barriers between different crystals impede the flow of electrons and thus limit the cell's ability to convert sunlight into electricity. No-crystal or amorphous silicon solar cells maintain silicon in a liquidlike state, even though they have all the outward qualities of being solid (as is also true for glass), and hence the atoms do not combine into a crystalline structure. Amorphous silicon cells promise to be the most inexpensive to manufacture, since they can easily be produced in thin films, but their conversion efficiency is lower than that of single and polycrystalline cells.

A silicon atom has 14 protons and 14 electrons configured in three shells, with four electrons in the outer or valence shell. Silicon atoms generally bond together into a three-dimensional structure called a *crystal lattice*. Each of the four outer or valence electrons is shared with four other silicon atoms to complete each atom's outer shell and create a very stable formation, as shown in Figure 2-2.

At 0°K or −273°C, an unilluminated crystal would have no free electrons or "holes" (space left by an absent electron in the bonded structure)

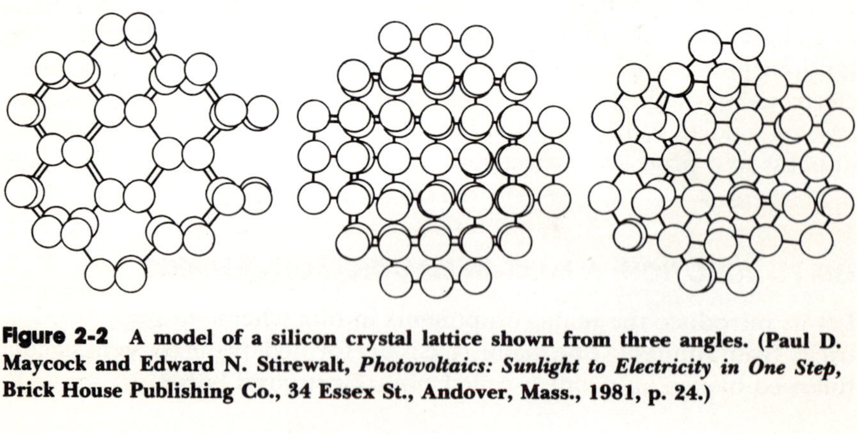

Figure 2-2 **A model of a silicon crystal lattice shown from three angles. (Paul D. Maycock and Edward N. Stirewalt, *Photovoltaics: Sunlight to Electricity in One Step*, Brick House Publishing Co., 34 Essex St., Andover, Mass., 1981, p. 24.)**

because there would be no energy available to liberate electrons from the crystal lattice. As the temperature rises, however, thermal energy increases the molecular vibration and is capable of knocking electrons out of their valence band into the conduction band. Holes are created wherever an electron has departed. Light similarly liberates electrons. Nonetheless, these free electrons have no place to go and invariably fall back into a hole, releasing energy as they sink. If a voltage, such as that produced by a battery, is put across such a piece of pure silicon, this electromotive force can cause the free electrons to hop from one hole or atomic orbit to the next as they migrate toward the positive terminal, while the holes created by the missing electrons move toward the negative terminal.

If we were to stick a pure silicon wafer out in the sun, it would heat up admirably, but little else would occur. So far the plot is uneventful, if not dull. Our scheme requires us to create a voltage or electromotive force *inside* the silicon that will separate the electrons and holes and send them through a circuit. This accomplishment involves some trickery. At this point, we decide to introduce two antagonists into the silicon crystal; one is the phosphorus atom, and the other is the boron atom. The phosphorus atom is attractive to our scheme because it has five electrons in its outer shell, and the boron atom is chosen because it just has three.

Let us slice our piece of pure silicon in half and insert some phosphorus atoms into the silicon crystal lattice at a ratio of about 1:1,000,000. Similarly, let us insert boron atoms into the other piece. The process of intentionally inserting impurities, called *doping,* creates free electrons in the phosphorus-doped piece of silicon and holes or gaps in the crystal bonds of the boron-doped piece. Every silicon atom covets four other silicon atoms to share its four valence electrons. When a silicon atom is forced to accept a phosphorus atom instead, the silicon atom's outer shell will be complete, but the fifth valence electron from the phosphorus atom has no strong allegiance to the partnership and is easily persuaded to wander away. When the silicon structure is forced to accept a boron atom, on the other hand, it must leave the outer shell of one of its atoms incomplete, with only seven electrons. Boron has only three valence electrons and thus deprives one silicon atom of an electron to share. The absence of an electron in the crystal lattice creates a hole, which is an open invitation to any eligible electron that wishes to fill it. Let us call our phosphorus-doped silicon *N-type material* for the word "negative," since it has free electrons. Then let us call our boron-doped silicon *P-type material,* since it has many holes that act like "positive" charges by attracting electrons. Notice how silicon, boron, and phosphorus atoms arrange themselves in P- and N-type material, as shown in Figure 2-3.

If we were to stick our two pieces of differently doped silicon out in

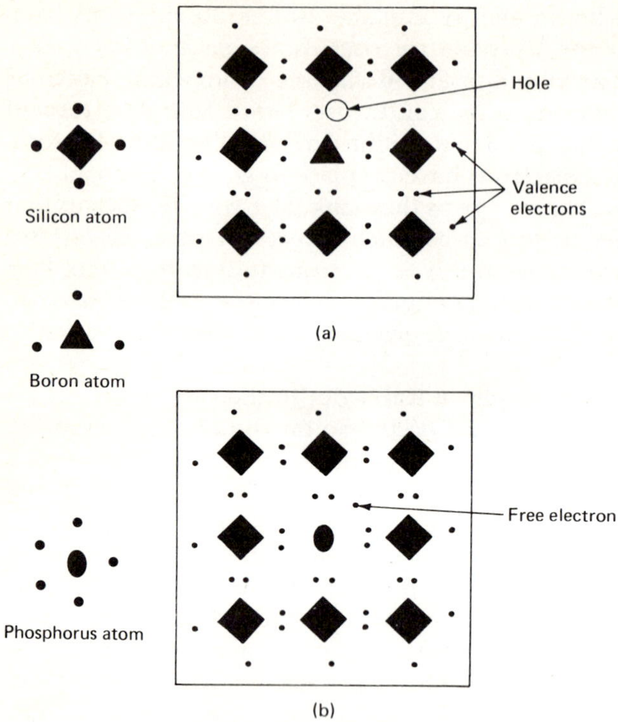

Figure 2-3 Silicon doped with boron becomes (*a*) P-type material, while silicon doped with phosphorus becomes (*b*) N-type material.

the sun, they would get warm, but little else could be expected. The drama is still devoid of dynamics. As in most stories, however, diverse bodies have a tendency to be attracted to each other. This tale is no exception, for we shall now fuse these two alternately doped pieces of silicon together and watch their chemistry interact. As expected, some of the free electrons in the N material are drawn into holes in the P material. Our fused piece of P- and N-type materials, aptly called a *PN junction,* is the fundamental building block of a solar cell.

Let us examine the two forces acting on a free electron in the N-type material. This electron belongs to a phosphorus atom and has a proton counterbalancing it in the phosphorus nucleus. The tight structure of four silicon atoms, however, weakens the phosphorus nucleus's grip on this fifth electron. On the other side of the PN junction, we have a boron atom bonded with four silicon atoms, creating a hole that desperately wants an electron to complete its formation. Near the junction of the two materials, the P-material hole has more power to draw electrons than does the phosphorus nucleus. Electrons wander from the N to the P material and fill the available holes, as shown in Figure 2-4, creating an area around the junction called the *depletion region.* This depletion region can grow to only a certain size before it builds up a barrier. The unneutralized area of the N material that has lost its electrons and the adjacent area of the P material that has had its holes filled are no longer conducive to the flow of electrons. A balance is achieved between the attraction for electrons of the P-material holes and the insulating barrier built up by the movement of electrons and holes across the junction. When the unneutralized depletion region is large enough to prevent electrons from reaching the holes on the other side, the tide across the junction diminishes. The width of the depletion region depends on the dopant concentration, but usually is less than 0.0001 cm.

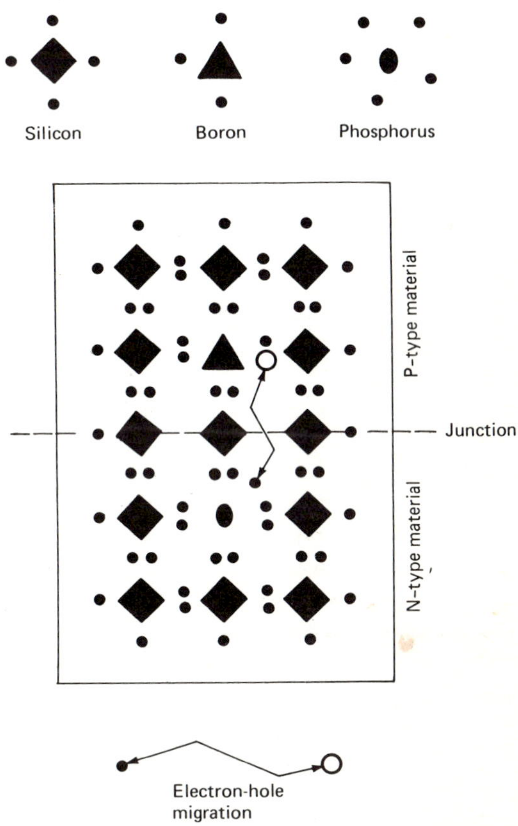

Figure 2-4 Electron and hole migrations across a PN junction.

Before the two pieces of differently doped silicon were fused together, the N and P materials were of neutral charge; the number of protons equaled the number of electrons. Even though one material has available electrons and the other has available holes, they both consist of neutral atoms that have their normal number of electrons. The act of fusing the P and N material together and the resulting migration of electrons from N to P material and the reverse flow of holes suddenly creates an imbalance. The N material has lost some of its valence electrons and hence acquires a net positive charge, and the P material has gained some electrons and thus receives a net negative charge. A small voltage, called the *barrier potential,* is created across the junction that prevents the further flow of electrons from N to P material and that sweeps all the free electrons and holes out of the depletion region. We have engineered a permanent electromagnetic force that is built into a block of material.

Up to now we have had little problem creating free electrons with thermal and light energy, but a force or direction has been missing. Our free electrons have been randomly tossing about, appearing and disappearing. The PN junction–induced voltage could be the force we need to send electrons traveling in a productive direction. How might we cleverly harness this imbalance that we have created in the fine structure of matter?

We have to account for so many agents flowing back and forth across the border that it might be appropriate at this time to give them some names. There are a majority of electrons in the N material, so let us call them the *N majority carriers.* Similarly, there are a majority of holes in the P material, so let us call them *P majority carriers.* Wherever there is a majority, there usually is its counterpart, the minority. Unlikely as it may seem, there are some holes in the N material, created when the thermal and light energy knocks electrons out of their shells. These few oddities we shall call the *N minority carriers.* Also, there are some free electrons in the P material, created by that same thermal and light energy. As you guessed, we will call these electrons the *P minority carriers.* Figure 2-5 depicts the flow of majority and minority carriers in a PN junction.

The population of majority carriers in both P and N material is determined by the concentration of boron and phosphorus doping atoms and is not significantly influenced by light or thermal excitation. On the other hand, the number of minority carriers is directly related to the amount of heat and light energy that the PN junction is exposed to. Minority carriers are the vehicles of solar power in a solar cell.

When a PN junction sits in the dark at room temperature, a certain number of free electrons and holes pop up in a peripheral area of the cell not especially near the junction and jump around in a helter-skelter fashion before eventually recombining and falling back into the crystal

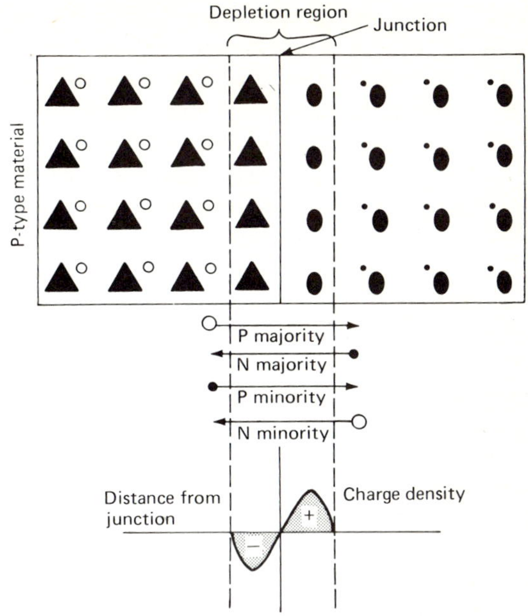

Figure 2-5 **Majority and minority carrier flow across a PN junction.**

lattice. The free electrons and holes that appear near the junction in the depletion region, however, meet a different fate. These electrons and holes fall under the influence of the barrier potential and are swept out of the depletion region. The electron is repelled by the P material because that material already has a surplus of electrons, and thus flows into the N material; the hole moves in the opposite direction. As you have surely noticed, the direction in which these thermally generated electrons are flowing is opposite to the electron flow caused by the P material's desire for an electron to fill its holes. In an unilluminated PN junction, the majority and minority currents are equal and balance each other out. We could not hook up wires to such a junction and expect to get any current from it, because it is in thermal equilibrium with its surroundings.

Of course, our PN junction by itself will not produce electricity. Electricity requires both a voltage and a current, and while there are plenty of free electrons roaming through our little block, they flow, as we have seen, in such a way as to exactly cancel each other out. If we connected wires to the ends of our crystal, no power would be produced. But if the flow of electrons were not in balance, so that the embedded voltage had a few extra electrons to push, we might be able to get electricity out of our PN junction. It should come as no surprise, then, to see that the addition of light, a form of energy which naturally produces free electrons in any material it hits, will provide the kick to get our electric power flowing.

Let us take a small, square piece of N-type silicon about 0.1 cm thick and 5 cm on a side, and attach it to a metal base. On the top of this assembly we will place a tiny cover of P-type material, a few thousandths of a centimeter thick, to form a broad, flat PN junction. Next we will place on the surface of our P-type silicon a thin wire mesh, carefully designed so that it is thick enough to collect electrons from every corner of the crystal, and yet sparse enough so that it blocks little of the incoming light. Single-crystal silicon ingots, wafers, and solar cells are shown in Figure 2-7.

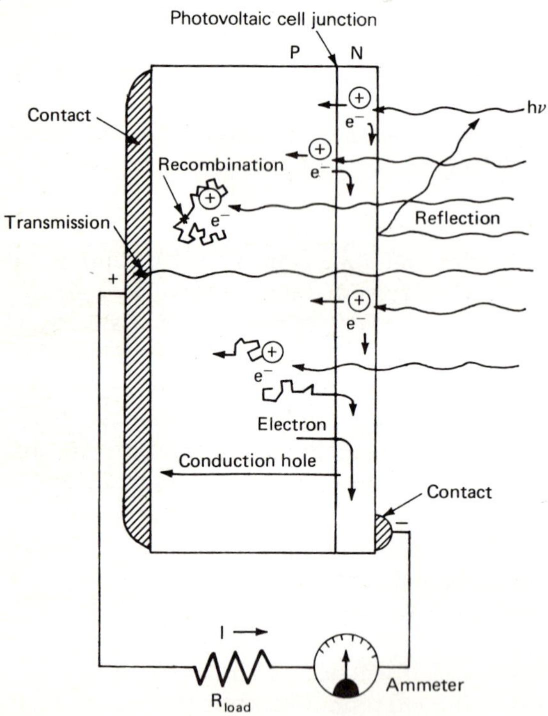

Figure 2-6 **Photovoltaic-generated current flows through a circuit. (From: Joseph Merrigan, *Sunlight to Electricity*, M.I.T. Press, Cambridge, Mass., © 1975, p. 35.)**

If we were to put our cell in the dark and put an ammeter across the back metal plate and the front metal mesh, we would, as we have seen, measure nothing. Now suppose that a single particle of light, a photon, descends into our cell. This tiny bundle of energy may strike an electron somewhere in the valence band. We have seen how such an electron can occasionally absorb thermal energy and promote itself upward. The same result can be produced by light energy. If a photon whose energy is greater than that of the energy gap strikes a valence electron, that electron will absorb the photon and jump into the conduction band.

If such an event occurs in the depths of the N- or P-type material, the newly created free electron and hole will quickly be absorbed by other electrons and holes, resulting in *recombination.* But if a photon strikes a valence electron circling an atom in the depletion region, the electron and hole will be forced apart by the electric field and pushed toward their respective N and P materials. One electron and one hole will have been added to our previously balanced recombination and generation currents. We will have a net current of one electron (or one hole), which the voltage will promptly shove into any wire attached to the end of the crystal. The ammeter will record a current corresponding to one electron.

Of course, even on the cloudiest days, trillions of photons beam down from the sky. As these photons cascade down into our photovoltaic cell,

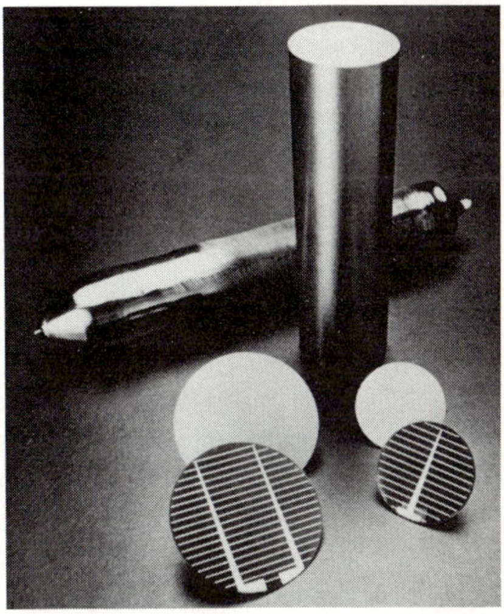

Figure 2-7 Single-crystal silicon ingots, wafers, and solar cells. (Jet Propulsion Laboratory.)

they knock trillions of electrons out of the valence band, creating trillions of electron-hole pairs. Many of those pairs do nothing but quickly recombine, heating up the crystal in the process, but those pairs created in the depletion region are quickly swept apart and propelled into the metal base or wire mesh by the voltage across the PN junction. The ammeter will record a current of several amperes, as shown in Figure 2-6.

Thus we have succeeded in producing a device that gives us both a voltage and a current. We can attach the wires from the metal base and mesh to a load, such as a motor, a heater, or a radio, and the PN junction of our photovoltaic cell will force a current of light-created electrons through them.

PHOTOVOLTAIC CONVERSION EFFICIENCY

The efficiency with which a photovoltaic cell turns sunlight into electricity is only about 15%. The theoretical maximum efficiency is about 25%. It might be interesting to use our understanding of PN junctions to figure out what happens to the other 75%. The sunlight striking the earth's atmosphere carries about 1350 watts of energy per square meter. By the time this light reaches the surface of the earth, it carries about 1000 peak watts per square meter. The photons which make up this light have a range of different energies, as shown in Figure 2-8. The majority of the inefficiency in a photovoltaic cell is caused by the energy of the sun's photons being so spread out.

The perfect photon would have an energy exactly equal to the band gap of silicon. A photon with less than this amount of energy will not be able to knock a valence electron all the way to the conduction band. Since an electron can be in either the conduction band or the valence band, but never in between, photons with insufficient energy pass straight through the silicon crystal with no effect. This wastes about 24% of the sunlight.

Overly energetic photons are equally wasteful. Once an electron is in the conduction band, it has little use for extra energy, and since a photon must be absorbed completely or not at all, the extra energy a valence electron picks up, above and beyond what it needs to get to the conduction band, does nothing but make the crystal hot. This wastes another 32% of the incoming energy.

These two inefficiencies are inherent in nature, and as long as we use silicon for our solar cells, there is not too much we can do about them. Next we come to a whole series of little inefficiencies (our list is by no

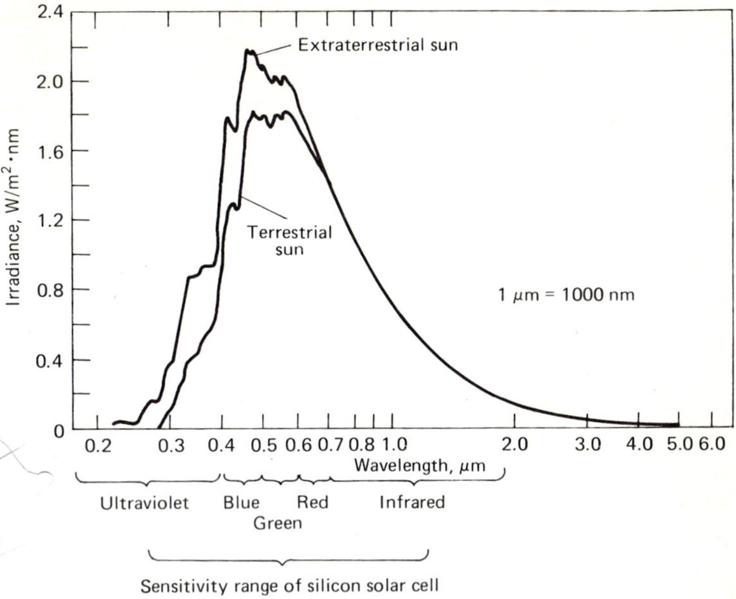

Figure 2-8 **The range of light waves that are constructively absorbed by solar cells.**

means exhaustive) that are more under our control. While we can never eliminate them, clever engineering can limit their effect.

The first of these problems is that light striking a shiny surface like that of crystalline silicon has a tendency to be reflected. To keep photons going into the crystal instead of bouncing off, manufacturers coat their cells with an antireflective coating. Still, about 3% of the energy is lost through reflection.

Impurities in the crystal are also a problem, as they tend to hinder the motion of the mobile electrons and holes. This is a serious difficulty because the electrons and holes generated in the depletion region have only a limited lifetime before they run into another hole or electron and are absorbed. These carriers are not safe until they have reached their respective N or P region. If the time these particles need to move out of the depletion region is increased because they keep bumping into impurities in the crystal, many of the carriers will be lost, and the efficiency of our cell will go down. It is vital, of course, to keep the phosphorus and boron we use for doping, but ideally we would like to be rid of any other imperfections in the silicon crystal. The purer we make the crystal, however, the more expensive it gets, and so we have a tradeoff between purity, efficiency, and price. Different manufacturers handle this trade-off in different ways, and this area is one of the most fertile for technological advance in the solar industry.

Another problem is that a certain number of photons are absorbed in parts of the crystal far from the depletion layer. These photons create electron-hole pairs which do nothing but immediately recombine, leaving only a little heat as their legacy. Since the region of the crystal near the PN junction is the only part that actually contributes to the generation of electricity, ideally we would like a photovoltaic cell to be a few thousandths of a centimeter thick. Again we have a manufacturing trade-off and an area ripe for technological improvements. At present it is impossible to make a crystal of sufficient purity that is thinner than human hair—and even harder to do it cheaply.

These two inefficiencies, impurities in the crystal and absorption away from the depletion region, account for about 20% of theoretically unavoidable loss.

Several times we have seen that many of the photons that enter a solar cell, if they don't release their energy to produce electricity, will release it to produce heat. We will find in later chapters that the hotter a solar cell is, the less power it produces. It might be interesting to examine why. Remember, when we first made our PN junction, the depletion layer and its voltage grew until the recombination current was exactly cancelled by the generation current. The generation current is caused by electrons being knocked out of the valence band by the thermal motion of the atoms of the crystal. Now, as one might imagine, the hotter the crystal, the faster these atoms will vibrate, and the more electron-hole pairs will be created. In other words, the hotter the crystal, the bigger the generation current. But the generation current was the limiting factor on the growth of the depletion layer. The bigger this current, the smaller this layer and its associated voltage. A smaller depletion layer is a smaller target for incoming photons, and so fewer will be absorbed there. This means less light-generated current and, combined with the smaller voltage, less photovoltaic power.* That is why a photovoltaic cell works best on a bright yet chilly day.

OTHER TYPES OF PHOTOVOLTAIC CELLS

Up until now, we have considered only a single crystal of silicon as the basis of our photovoltaic cell. We took this tack because at present cells

*"Wait," you might say, "isn't the current generated by the photons in effect just an increase in generation current as well?" It definitely is, but to say that light decreases the cell's efficiency is meaningless, because without light the cell produces no power. The point is that the heat is useless energy; it decreases the size of the voltage but gives us nothing in return.

based on the single crystal are the most commonly manufactured and the most thoroughly understood of any of the myriad ways in which it is possible to produce photovoltaic power. To conclude this chapter, we will give a brief review of some variations, ranging from factory prototypes ready to test the market to ideas that may be little more than laboratory notes.

Polycrystalline silicon is a block of material which, instead of being a single crystal, is made up of many tiny crystals pieced together. Several manufacturers have pioneered processes for mass-producing inexpensive polycrystalline cells, taking advantage of the fact that it is inherently easier to grow little crystals than big ones. A disadvantage of polycrystalline cells is that the boundaries between the tiny crystals tend to impede the flow of holes and electrons through the material. These boundaries either act as barriers which slow carrier motion or actually provide a path for electrical shorts across the cell. In either case, the cell's efficiency suffers, although several techniques for coping with these problems have proved successful enough for 10% efficient polycrystalline cells to be produced.

To produce single and polycrystalline cells, we must grow the crystalline material out of a pool of molten silicon. Once grown, the material must be cut into the thin wafers required for photovoltaics. We have seen, however, that the only active part of the photovoltaic cell is the depletion region, a few millionths of a centimeter thick. Since it is impossible to cut anything this thin,* much of the material in a photovoltaic cell goes to waste. One way around this is to abandon the crystalline state altogether and, using such techniques as the condensation of gaseous silicon, make cells whose thickness can be measured in numbers of atomic layers. The atoms in such thin films of silicon are arranged in completely random fashion, and the cell is called an *amorphous thin-film cell.* Though these cells are inexpensive, abandoning the crystal structure wreaks havoc on their efficiency. About 10% is the best that has ever been achieved. The hope is, however, that this figure can be greatly improved. Understanding the properties of amorphous materials is much more difficult than understanding those of crystals, simply because the regular arrays of atoms in the latter are much more amenable to mathematical description than the random distributions of atoms in the former. So far, amorphous cells have been designed with little more guidance than trial and error, and perhaps, as researchers develop a better understanding of amorphous-cell behavior, the designers of these cells can greatly enhance their performance.

The photovoltaic process is not confined to silicon. Though germa-

*Some manufacturers can grow ribbons directly without the need for cutting, but these ribbons are still not nearly as thin as the depletion layer.

nium is the only other element that is a good semiconductor, there are various chemical compounds whose band gaps and other electrical properties make them suitable for converting light into electricity. Each of these materials has its own advantages and disadvantages, but among the most commonly used and studied are copper-cadmium sulfide and gallium arsenide.

Copper-cadmium sulfide is usually used in thin-film cells constructed by placing a minute layer of Cu_2S on a slightly thicker layer of CdS. The cells seem inexpensive to produce and have demonstrated efficiencies up to 9%. The chemical instability of Cu_2S is a problem, however, and the cells must be carefully sealed off from the environment so that they will not be destroyed.

Gallium arsenide can be successfully used in amorphous thin-film, polycrystalline, and single-crystal cells. In this last form, it has achieved the highest efficiency yet demonstrated by a photovoltaic cell, about 22%. This high efficiency makes it an ideal material for use in sunlight concentrator systems, where it shows much promise. Both the elements involved in the cell are problems, however, the former, gallium, being scarce, and the latter, arsenic, being toxic.

There is no shortage of materials and designs which with suitable care can be made to turn light into electricity. The tests of economics and reliability will be the real constraints. It is not yet clear what methods—flat-plate collectors, concentrators, or thin-film cells made out of silicon (crystalline, polycrystalline, or amorphous), copper-cadmium sulfide, gallium arsenide, or something entirely different—will ultimately prove the most effective. For the present, however, crystalline silicon flat-plate collectors are the foundation of the photovoltaic industry, and in the chapters that follow we will learn how silicon PN junctions can best be gathered into arrays, pointed at the sun, and used to provide electric power.

3

SOLAR ENERGY

The designer of a photovoltaic system must be able to ascertain the amount of sunlight that will exist per given area at different locations and times. A combination of predictable and less predictable considerations are involved in determining this expected solar energy. The sun's intensity on earth is governed by two factors: the amount of atmosphere the sun's rays must pass through and the angle with which its rays strike a surface. Most of the variables that determine these two factors vary in a consistent and predictable fashion. The sun's position in the sky can be determined from the time of day and year and the latitude of a site in question. The slope and position of a tilted surface on earth are controlled by the operator. To the chagrin of many people, however, nature introduces a less predictable variable, namely weather. Despite the cyclical variations during the four seasons, weather patterns involving temperature, cloud cover, wind, and precipitation can be quite whimsical. Long-term weather prognosis can only be approximated by taking weather data for any location in question and averaging the same day's or month's weather record over many years.

THE SUNLIGHT EXPERIENCED ON EARTH

The solar power received directly outside the earth's atmosphere is continuous at a rate of about 1350 watts per square meter (minor perturbations resulting from sunspots and eccentricities in the earth's orbit cause this value to fluctuate about ±3%). More specifically, a 1-m^2 area perpendicular to the sun's rays and located about 55 kilometers above the earth's surface would receive energy at the rate of 1350 W·s every second. This insolation figure is stable enough to be called the *solar constant.*

The solar power experienced on the earth's surface, however, is neither continuous nor as plentiful. On earth, the solar-power level fluctuates with the rotation of the earth on its axis and the rotation of the earth around the sun; during different times of the day and year, the sun's intensity is changed by the varying amounts of atmosphere solar radiation must pass through and the changing incident angle at which the light strikes any surface. The peak solar power experienced on a horizontal surface near sea level is about 1000 watts per square meter when the sun is at its apex in the sky on a sunny day. This figure is often called *one sun* and is the power level at which solar cells usually are rated; photovoltaic cells are said to produce so many *peak watts* per given area.

The sun's radiation interacts with the earth's atmosphere in four different ways. As sunlight passes through the atmosphere, some of the sun's radiation is reflected back into outer space, while another portion is absorbed or scattered by air molecules, water vapor, and dust. The light that finally reaches the life-bearing region on the earth's surface is either direct-beam or diffuse.

The sunlight received by a surface on earth can be divided into three different types: direct, diffuse, and reflected, as shown in Figure 3-1. Diffuse sunlight approaches a surface from all unobstructed angles, while direct-beam rays strike the surface from only one angle. For example, when the sun's direct-beam light is blocked by some object on earth, such as a tree, the shaded area does not plunge into complete darkness because light diffused by the atmosphere illuminates an area from all sides. However, when the sun's direct-beam light is blocked by the moon during a solar eclipse, a near darkness falls over the earth because there is negligible atmosphere between the sun and the moon to provide alternative paths by which the light can reach us. In addition to diffuse and direct light, an angled surface can also receive reflected light from the ground or other appropriately positioned surfaces.

The proportions of direct, diffuse, and reflected sunlight that a particular surface will receive depend on variables that are in continuous flux. On a cloudy day, all radiation is diffuse. On a sunny winter day at a low-elevation, humid-climate site, the ratio of diffuse to total radiation could be as high as 0.7. In contrast, on a sunny summer day at a high-altitude arid-climate site, the ratio might be as low as 0.2. More light is reflected by pale than by dark colors. Light reflection often increases in

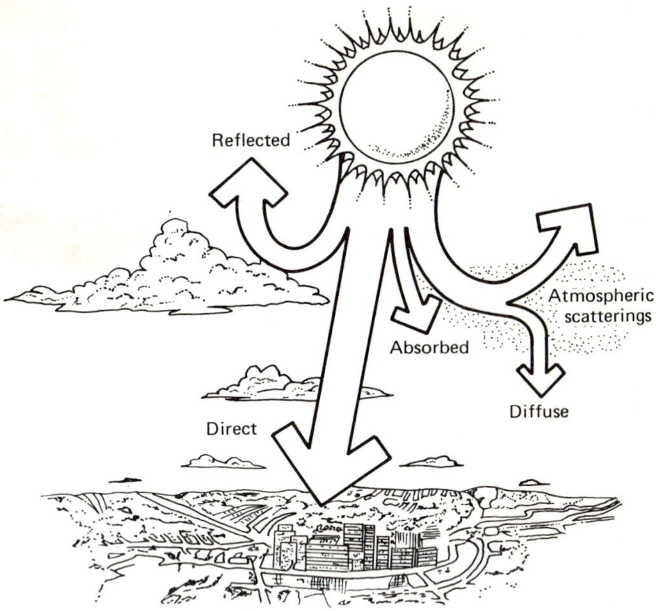

Figure 3-1 **Sunlight and the earth's atmosphere. (From: Edward Mazria, *Passive Solar Energy Book,* Rodale Press, Emmaus, Pa., © 1979.)**

the winter when there is snow on the ground and decreases in the summer when the ground is earth- or grass-colored. Snow mirrors about 70% to 80% of the light it receives, while a grass field reflects only about 15% to 20%.

SURFACE ORIENTATION AND SOLAR INTENSITY

The orientation of a surface on earth is defined by two angles, the surface tilt or slope angle and the surface azimuth angle, as depicted in Figure 3-2. The tilt angle indicates how far up from the horizontal a surface is slanted, while the azimuth angle denotes how the surface is positioned relative to the true north-south and east-west coordinates (due south represents an azimuth angle of 0°, due east is −90°, north is 180°, and west is +90°). A horizontal surface with a zero slope receives the maximum diffuse and minimum reflected light possible. As a south-facing surface is tilted up from the horizontal, the amount of diffuse light decreases while the proportion of light reflected off the ground increases.

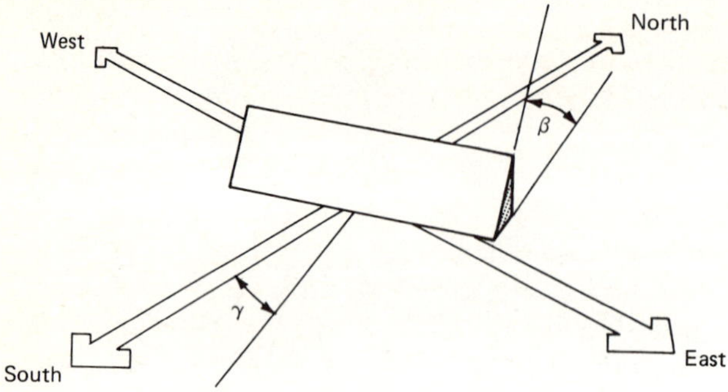

β = surface slope or tilt angle
$0 \leq \beta \leq 180°$
($\beta > 90°$ means that surface is facing downward)

γ = surface azimuth angle
$-180° \leq \gamma \leq +180°$
(east is negative and west is positive)

Figure 3-2 Orientation angles of a surface on earth.

The angle at which the sunlight passes through the atmosphere and strikes a surface determines the intensity of the energy received. If the sun is directly overhead, its radiation has a far shorter path through the obstructing atmosphere than when it is situated at a low angle from the horizon. The lower the sun's angle from the horizon, the more atmosphere its light has to traverse, and consequently the lower its intensity when it finally reaches the earth.

In addition to and irrespective of the atmospheric effect, any light beam becomes less concentrated, and therefore less intense, as its angle with a surface becomes smaller. A circular beam of light that is perpendicular to a surface illuminates a round area. As the angle of the light beam relative to the surface decreases from 90°, the lit area becomes elliptical and larger. The same quantity of light spread out over a bigger area results in less light per unit area.

LOCATING THE SUN'S POSITION

The earth's daily rotation on its axis and the annual rotation of the tilted earth around the sun both affect the angle at which sunlight passes through the atmosphere as seen from any one point on the earth. The position of a site on earth with respect to the sun is determined by two continuously changing angles—namely, the sun's hour and declination angles—and by one fixed angle that specifies a site's location on earth–namely, the latitude.

Hour Angle

The sun's *hour angle* for a particular location depends on the momentary position of the earth in its axial rotation. Since the earth makes a complete 360° rotation in 24 hours, the hour angle changes 15° every hour. The hour angle is measured from the local meridian, or the sun's highest point in the sky at solar noon, with angles between sunrise and solar noon being positive and angles after noon being negative.

The local standard time for solar noon is not necessarily 12:00 noon. Eccentricities in the earth's orbit and the fact that the earth is both rotating on its axis and revolving around the sun simultaneously cause solar time to vary from local standard time by a factor that is computed using the equation of time. The earth is divided up into 24 one-hour longitudinal sections called *time zones,* with the standard time in each zone being based on the solar time for the longitude in the center of that particular zone. The local standard time for locations not on the center longitude can vary from the solar time by as much as half an hour. The local standard time for solar noon at different longitudes and times of year can be computed using Equations (3-1) and (3-2).

$$\text{LA} = [(\text{local longitude}) - (\text{longitude of local time meridian})/15] \quad (3\text{-}1)$$

$$\text{Local standard time at solar noon} = 12 + \text{Eot} + \text{LA} \quad (3\text{-}2)$$

The Eot (equation of time) figure is obtained by looking up the month in question on the graph in Figure 3-3. The longitude of the local time

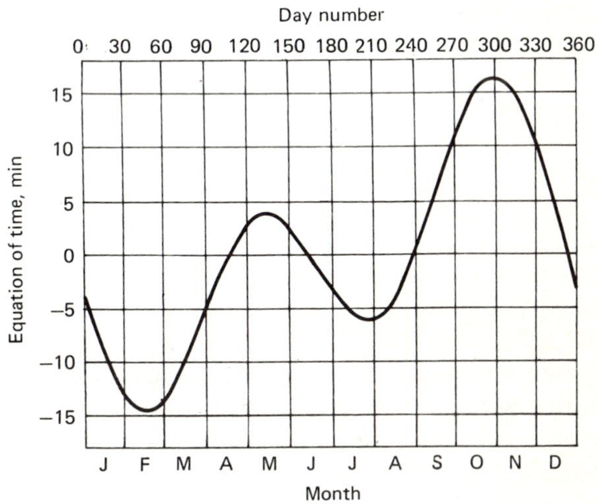

Figure 3-3 **The solar time correction factor versus time of year.**

meridian is obtained from an atlas with the time zones marked on its map. In the Northern Hemisphere, local standard time will generally be ahead 1 hour between April and October because of daylight savings, and this hour must be subtracted after any of the above calculations are made.

Declination Angle

The sun's declination angle is the angular position of the sun at its highest point in the sky with respect to the plane of the equator; it depends on the momentary position of the earth in its revolution around the sun. Changes in the declination angle are caused by a simple fact: The earth's axial tilt of 23.45° remains constant and in the same direction during the earth's entire orbit about the sun. In the Northern Hemisphere, the declination angle reaches its most northern and positive peak of +23.45° on June 21st (the summer solstice) and drops to its most southerly and negative peak of −23.45° on December 21st (the winter solstice). In the Southern Hemisphere the converse is true. The equation for determining the sun's declination depends on the day of the year (1–365) and is shown in Equation (3-3) on page 62 and in Figure 3-15 on page 66.

When observing the sun from a location other than on the equator, as most of us do, the site's latitude must be taken into account. In the Northern Hemisphere, the angle from the horizon to the sun's position at its apex in the sky (commonly called the solar-noon altitude angle) can be calculated by finding the difference between the perpendicular of 90° and the latitude angle of the site, and then adding the declination for any day in question. For instance, if we are located at a latitude of 40°, the sun's highest altitude in the summer will be $90° - 40° + (+23.45°) = 73.45°$, and the sun's highest altitude in the winter will be $90° - 40° + (-23.45°) = 26.45°$.

The Sun Chart: Solar Altitude versus Azimuth

If we know the declination, the hour angle, and the site's latitude, we can calculate the sun's position in the sky at any moment in time. The aforementioned three angles indicate how a place on earth moves with respect to the sun. Even though the scientific developments of the fifteenth century informed us otherwise, the sun appears to be moving with respect to a stationary earth. The sun moves through the sky from east to west in an arc that is higher in the summer and lower in the winter. The sun's apparent position with respect to a stationary earth is measured by two angles: the solar azimuth and solar altitude angles, shown in Figure 3-4. The altitude angle measures the sun's angular

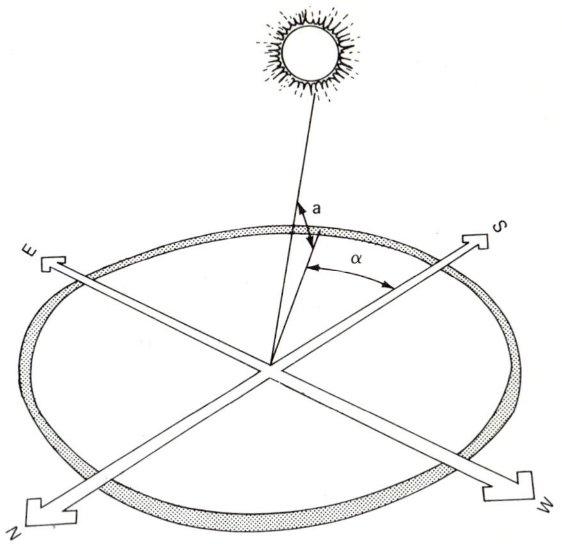

a = the sun's altitude angle (angular elevation above the horizon)
α = the sun's azimuth angle measured from local north-south meridian (angles east of due south are negative; angles to the west are positive)

Figure 3-4 Orientation of the sun in the sky.

distance from the horizon, while the solar azimuth angle measures the sun's angular distance from due south (the azimuth angles from sunrise to solar noon are negative, while the angles from solar noon to sunset are positive). The sun's azimuth and altitude are calculated from the latitude declination and hour angles of a place and time in question, using Equations (3-4) and (3-5).

$$\sin a = \sin \phi \sin \delta + \cos \phi \cos \delta \cos h \tag{3-4}$$

$$\sin \alpha = -\cos \delta \sin h / \cos a \tag{3-5}$$

where a = altitude of the sun (angular elevation above the horizon)

ϕ = latitude of the site

δ = declination of the sun

h = hour angle of the sun (angular distance from the meridian of the observer; east is negative, west positive)

α = azimuth angle of the sun (east is negative, west positive)

After the sun's azimuth and altitude angle are calculated for various times during the day, these values can be plotted against each other to produce a representative curve of the sun's path through the sky. By plotting such curves for various days of the year, the sun's path for different seasons can be compared. Sun charts showing the sun's path at different times of the year for different latitudes are shown in Figure 3-5*a* and *b*. (Solid lines on these charts represent the sun's path through the sky; dashed lines trace the time of day in local solar time.)

Sun charts are particularly useful for determining the shading effect of obstructions jutting up from the horizon, such as trees, buildings, and mountains. If the horizon profile (which in effect is the altitude and azimuth angles of objects standing on the horizon) can be transcribed onto our sun chart, we can determine during what times of the day or year a surface will be shaded. Shading during early morning and late afternoon hours has negligible effect, since the sun's intensity is quite low during these periods. However, if obstructions cast large shadows on a site between about 7:30 AM and 4:30 PM at any time of year, attempts should be made to remove the obstructions or relocate the site.

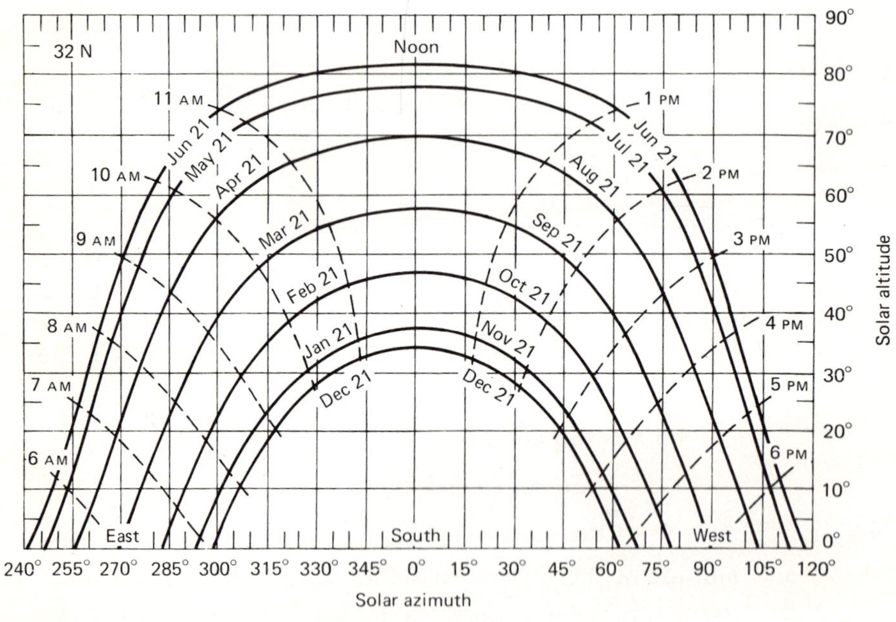

Figure 3-5 (*a*) **The sun's altitude versus azimuth angle for a latitude of 32° north. (Yvonne Howell and Justine A. Bereny, *Engineer's Guide to Solar Energy*, Solar Information Service, San Mateo, Calif., 1979, p. 30.)**

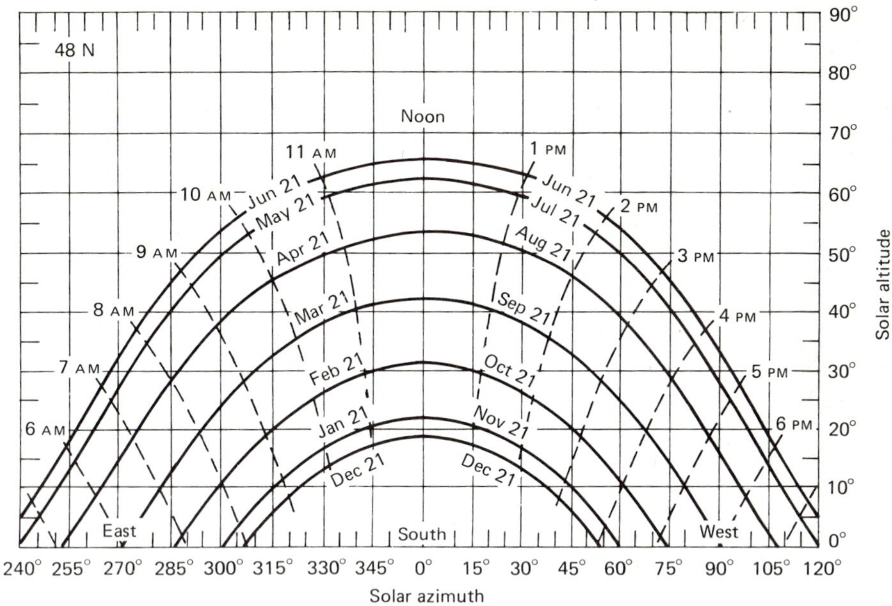

(b)

Figure 3-5 (*b*) **The sun's altitude versus azimuth angle for a latitutde of 48° north (Yvonne Howell and Justine A. Bereny, *Engineer's Guide to Solar Energy,* Solar Information Service, San Mateo, Calif., 1979, p. 32.)**

THE OPTIMUM SURFACE ORIENTATION

The amount of direct-beam solar energy that a surface receives is optimized by keeping the surface at right angles to the sun's direct radiation at all times. In most cases, maneuvering a surface is impractical. The cost of the energy and the mechanical assembly required to track the sun accurately and consistently each day generally outweighs the extra energy obtained by doing so when flat-plate solar-cell panels without concentrators are used. The next best proposition is to fix the surface's position so that the sun's angle on the plane is closest to the perpendicular most of the time.

In the Northern Hemisphere, the surface angle and position that faces the sun most directly throughout the year is directed due south and tilted from the horizontal at an angle equal to the site's latitude. North of the Equator, the sun's path is always located in the southern half of the sky unless the site latitude is below 23.5°, in which case the sun will spend most of its time in the southern sky. A surface must be positioned facing south to maximize its exposure to the sun over an

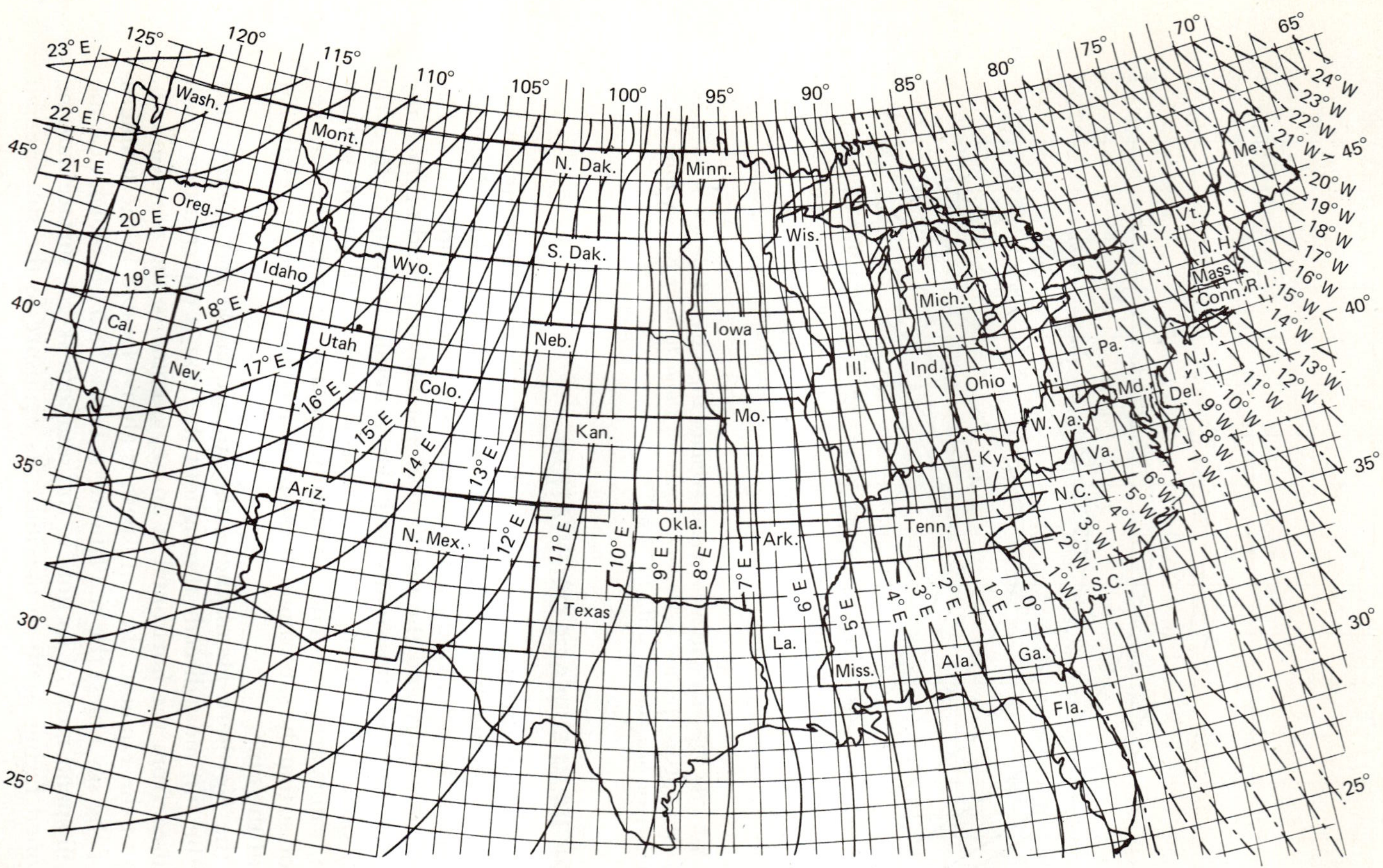

Figure 3-6 The isogonic chart of the United States showing the deviation between magnetic north and true north. For example, a magnetic compass reads about 15 degrees east of true north in Los Angeles, California. (U.S. Department of Commerce, Coast and Geodetic Survey, 1965.)

entire day. Due south can be located by using a compass and using the correction factor found in Figure 3-6 to correct for the discrepancy between true north and magnetic north, or by looking at the north/south shadow cast by a perpendicular stick at solar noon. To optimize the sun's energy, the surface must be angled up from the horizontal to face the sun. The optimum tilt angle over a full year must balance the summer maximum and winter minimum declination angle. The angle directly between the maximum and minimum altitude angles quite simply is equal to 90° minus the latitude angle of the site. The surface tilt angle that is most perpendicular to the sun's direct rays at this central solar altitude angle is equal to the site's latitude angle, as can be seen in Figure 3-7.

An area facing due south at a tilt angle that equals the site's latitude would obtain the optimum amount of direct-beam solar radiation over the entire year. If a designer wishes to maximize the solar energy received during the winter months, the surface tilt should approximately equal the latitude angle plus 11°, while the best orientation during the summer months is the site's latitude angle minus 11°. So far, however, we have neglected the diffuse and reflected components of solar radiation. Tilting a surface up from the horizontal decreases the diffuse radiation it can obtain and increases the reflection received from the ground. Diffuse and reflected light work at cross purposes here in a manner that complicates the picture.

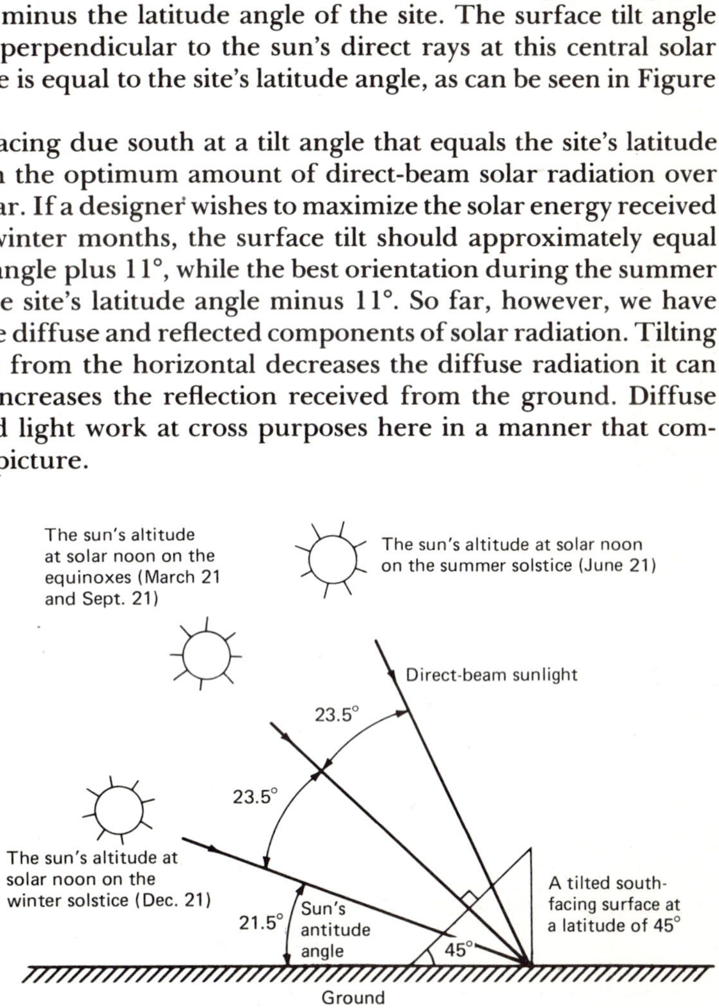

Figure 3-7 **The surface tilt angle that obtains the optimum amount of direct sunlight is equal to the site's latitude angle with the surface facing due south.**

The relative amounts of direct, diffuse, and reflected light must be known in order to make a precise determination of what surface tilt angle is optimum. Nonetheless, the surface tilt angle need not be especially precise; a deviation of as much as 10° from the optimum angle will produce a decrease in incident power of little more than 5% of the maximum, as is evident in the graphs shown in Figures 3-8 and 3-9. The annual optimum surface tilt angle is about 10% to 25% less than the latitude angle at sites with humid climates and therefore a considerable amount of diffuse light. In arid climates, the annual optimum tilt angle would be closer to the site's latitude angle.

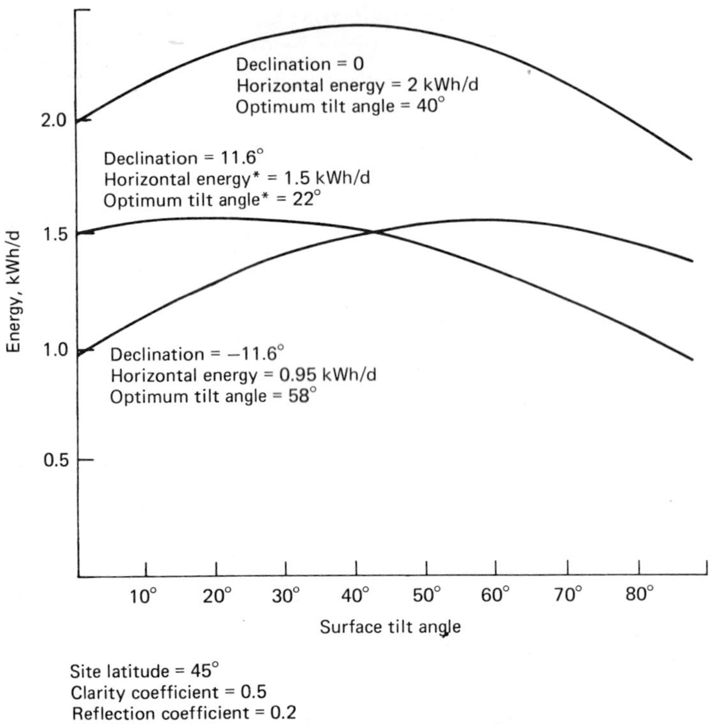

*These curves are based on Lui and Jordan's equations for converting energy on a horizontal surface to energy on a tilted surface. By their own admission, these equations are less accurate as the declination approaches the solstices (summer and winter). This optimum tilt angle is too low by at least 5°, and therefore the entire curve is somewhat suspect.

Figure 3-8 Solar energy versus south-facing surface tilt angle for three different times of the year.

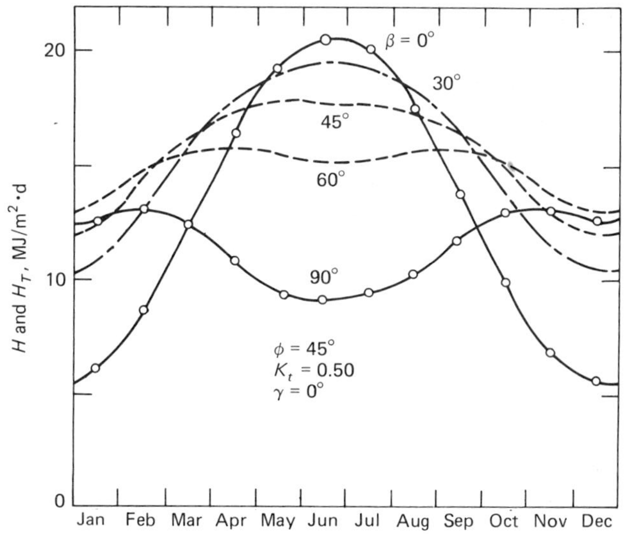

Figure 3-9 **Solar energy versus month of year for different south-facing surface tilt angles. (*a*, computed using Lui and Jordan's equation for converting insolation on a horizontal surface to insolation on a tilted surface; *b*, From: John A. Duffie and William A. Beckman, *Solar Engineering of Thermal Processes*, Wiley, New York, 1980, p. 98.)**

CALCULATING SOLAR ENERGY AND POWER

Solar power and solar energy can be represented in different ways: peak power, average daily power, and average daily energy. The distinction between power and energy must be fully understood. Energy is a state of matter that represents a "capacity to do work," while power indicates the rate at which an energy transaction is taking place. In other words, energy is an intangible quantity, while power is the rate at which that quantity can be, or is being, delivered or dissipated. Solar power is an instantaneous measurement of the rate at which energy is being received from the sun at any moment, while solar energy is measured over a period of time and is an accumulated capacity to do work.

Solar cells generally are rated in peak power or peak watts. The earth, however, experiences peak power for only about one hour around solar noon of each clear day. At latitudes between 0° and 50°, clear-day peak power on a latitude-tilted south-facing surface is about 950 watts per square meter plus or minus 15% throughout the year, with the power being on the low side in the winter and on the high side in the summer.

Instances of natural concentration can cause this figure to jump briefly to 1300 or 1400 watts per square meter when sunlight is both direct and reflected off clouds. Sunlight reflected off snow can also be significant.

Rating solar cells in peak power or peak watts can be misleading if the distinction between power and energy is not fully understood. As we continuously witness, the sun's power level rises from zero each morning and climbs to its maximum at solar noon, only to drop back to zero at sunset. Rating a solar cell in peak watts clearly puts it in its best light and is no indication of how it will perform over an entire day. A more representative solar-power rating would be the daily average power in clear weather, which is about 8/10 of one sun or 800 watts per square meter. An average power rating, nonetheless, leaves questions unanswered because the length of a day varies throughout the year. A comprehensive solar-cell rating would be the average energy output per day. Since the average daily energy varies in different parts of the country and on different days of the year, this figure would have to be qualified by referring to a particular site and time of year.

The average daily energy per unit area, the power level versus time per unit area, and the daytime temperature and precipitation are vital information needed to design a photovoltaic array. Measured insolation data automatically incorporate most of the variables involved in calculating solar energy. The length of the day, the height of the sun's path, the altitude and latitude of a site, the direct-beam and diffuse components of light, and the weather patterns are all taken into account when insolation data are measured. Determining the solar power versus time of day will indicate when and how long this solar energy is available. The daytime temperature affects the voltage and efficiency of a solar cell, and the precipitation indicates how often the cell will be covered with snow.

The major government agency responsible for collecting solar insolation data for the United States is the Department of Commerce Weather Bureau Climatological Center; for the world it is the World Meteorological Organization (WMO). Various universities and private concerns have established independent weather stations that can also be reliable sources for climatic data. The most common instruments used for measuring insolation utilize either the thermoelectric or the bimetallic expansion technique. Thermoelectric instruments (normally called *pyranometers*) can be accurate within 3% to 5% if properly operated, while bimetallic instruments can at best be expected to maintain an accuracy within 10% or more. Pyranometers are widely used in the United States, Europe, and Africa, while bimetallic expansion instruments are generally used in South America and Asia.

When assessing the reliability of insolation data, three issues must be considered. First, how accurate are the insolation-measuring instruments, and have they been maintained properly? The major causes of errors in data has been improper weather-measuring operation. Second, over how long a period have the insolation data been measured? Insolation should be measured at a site for at least 10 to 20 years in order to average out aberrant weather patterns (in the United States, weather measuring began to be widespread and reliable only during the early 1950s). Long-term changes in weather patterns resulting from astronomical considerations or human impact on the atmosphere will inevitably be overlooked by weather-measuring activities during the recent past. Third, how close is the proposed photovoltaic system site to the weather station? In most cases, a solar electric generator will be installed at a considerable distance from the nearest source of solar insolation data. In such instances, a careful comparison must be made of the differences in the terrain of the weather station and that of the solar installation. Differences in altitude, latitude, vegetation, and proximity to mountain ranges and bodies of water must be considered and accounted for.

Average Daily Solar Energy

Most daily solar energy data available for the past few decades have been measured on a horizontal surface. The energy-maximizing orientation (tilt and surface azimuth angle) of a surface is not considered in these data. In the Northern Hemisphere, it is fair to assume that a surface will be facing due south (with a surface azimuth angle = 0), since this position optimizes the energy received throughout the day any place north of the Equator. The optimal tilt angle over an entire year, however, depends on the latitude of a site and therefore can vary considerably. The horizontal energy data must be adjusted to energy expected on a tilted surface; a series of equations have been developed to perform this calculation.

The conversion of average daily energy per square meter on a horizontal surface to a tilted surface depends on the declination (time of year), the latitude of a site, and the tilt angle of a south-facing sloped surface. It is also necessary to know the relative proportions of direct, diffuse, and reflected light that a tilted surface would experience. The first stage in the conversion process involves determining the ratio between tilted and horizontal radiation. The final stage is to multiply this ratio by the average daily horizontal energy to provide an accurate value for the energy on a south-facing tilted surface. Let us go through these calculations step by step.

The relative proportion of diffuse to total insolation (assuming light reflection is zero) is calculated from the clarity coefficient K_t using the empirical formula found in Equation (3-9). The clarity coefficient is the ratio between the insolation experienced at a site and the insolation that would be experienced if the earth had no atmosphere; this coefficient is based on measured data and is given for different sites and months of the year in Appendix C. Since light reflection is assumed to be zero, the ratio of direct-beam to total insolation for a horizontal surface must be 1 minus the ratio of diffuse to total insolation.

The ratio of direct-beam insolation on a tilted surface to direct-beam insolation on the horizontal can be estimated (according to Lui and Jordan) by ignoring the effects of the atmosphere on sunlight and using Equation (3-8). The variables in this equation are the latitude, declination of the sun, surface tilt angle, and sunset hour angle of the horizontal and the tilted surface. The declination for any day can be found by using Equation (3-3) or by reading it off the graph in Figure 3-15. The sunset hour angles for the horizontal and for the tilted surface are computed using Equations (3-6) and (3-7). During the summer months, the sun may set on the tilted surface before it sets on the horizon (at these times the sun also would rise first on the horizon and then on the tilted surface). Consequently, Equation (3-7) requires that we take the minimum of two calculations.

The amount of light that is reflected off the ground is estimated by using a factor that ranges from 0 to 1. Common reflection coefficients for different materials are listed in Table 3-1. By visually comparing the reflections off different materials and using the list in Table 3-1, a reasonably accurate reflection coefficient can be assumed.

The ratio of average daily energy on the tilted surface to that on the horizontal surface is calculated using Equation (3-10). Notice how the equation is the sum of three different factors that indicate the relative proportions of direct-beam, diffuse, and reflected light. Multiplying the daily horizontal energy for the location and month under consideration by the ratio computed above will provide us with the information we

Table 3-1

Material		Coefficient
Snow	=	0.7–0.87
Concrete	=	0.31–0.33
Tar and gravel roofs	=	0.12–0.15
Asphalt paved roads	=	0.10–0.12

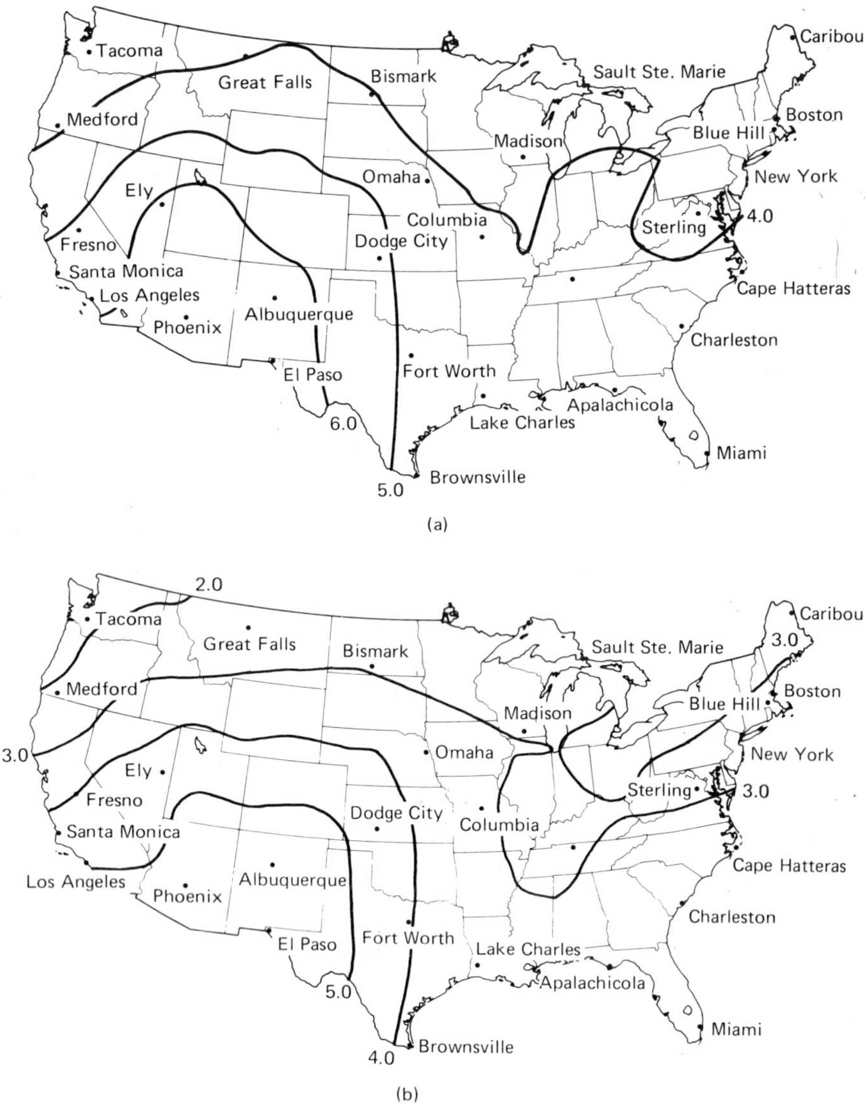

Figure 3-10 (*a*) Average daily availability of total solar radiation on a south-facing 45° tilted surface for the United States in the fall, in kWh/m². (*b*) Average daily availability of total solar radiation on a south-facing 45° tilted surface for the United States in the winter, in kWh/m². (*c*) Average daily availability of total solar radiation on a south-facing 45° tilted surface for the United States in the spring, in kWh/m². (*d*) Average daily availability of total solar radiation on a south-facing 45° tilted surface for the United States in the summer, in kWh/m². (Kreider, Jan F., Frank Kreith, *Solar Energy Handbook*, McGraw Hill Book Company, New York, 1981, pp. 2-35–2-37.)

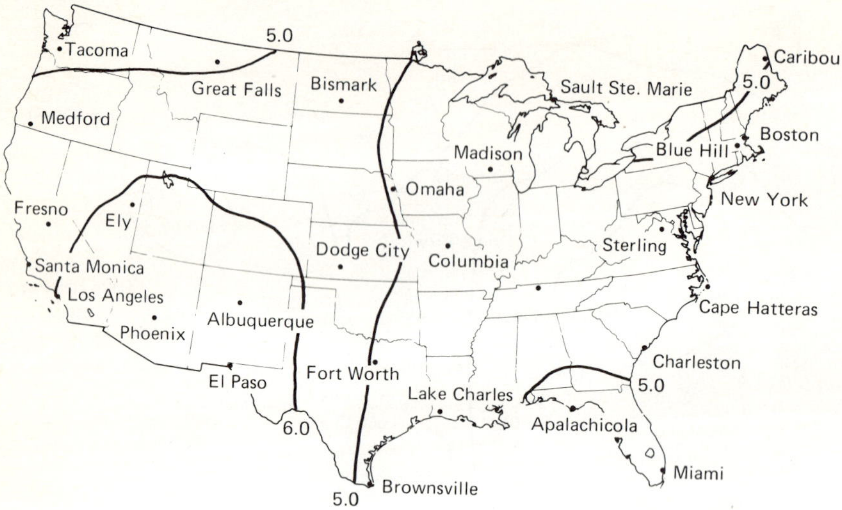

(c)

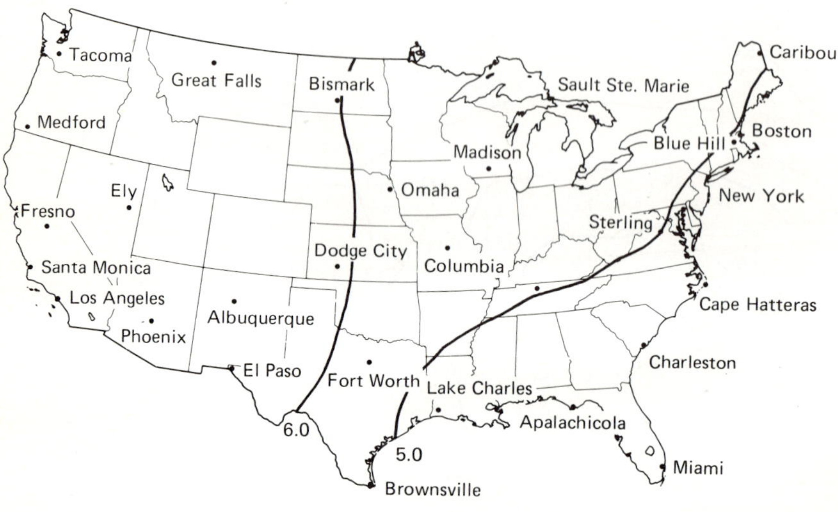

(d)

are looking for: average daily energy per square meter on a tilted surface for a specific site and time. Figure 3-10*a* to *d* and the tables in Appendix C summarize the solar energy available on a south-facing tilted surface at various locations in the United States. Figure 3-11 depicts the annual average horizontal insolation throughout the world.

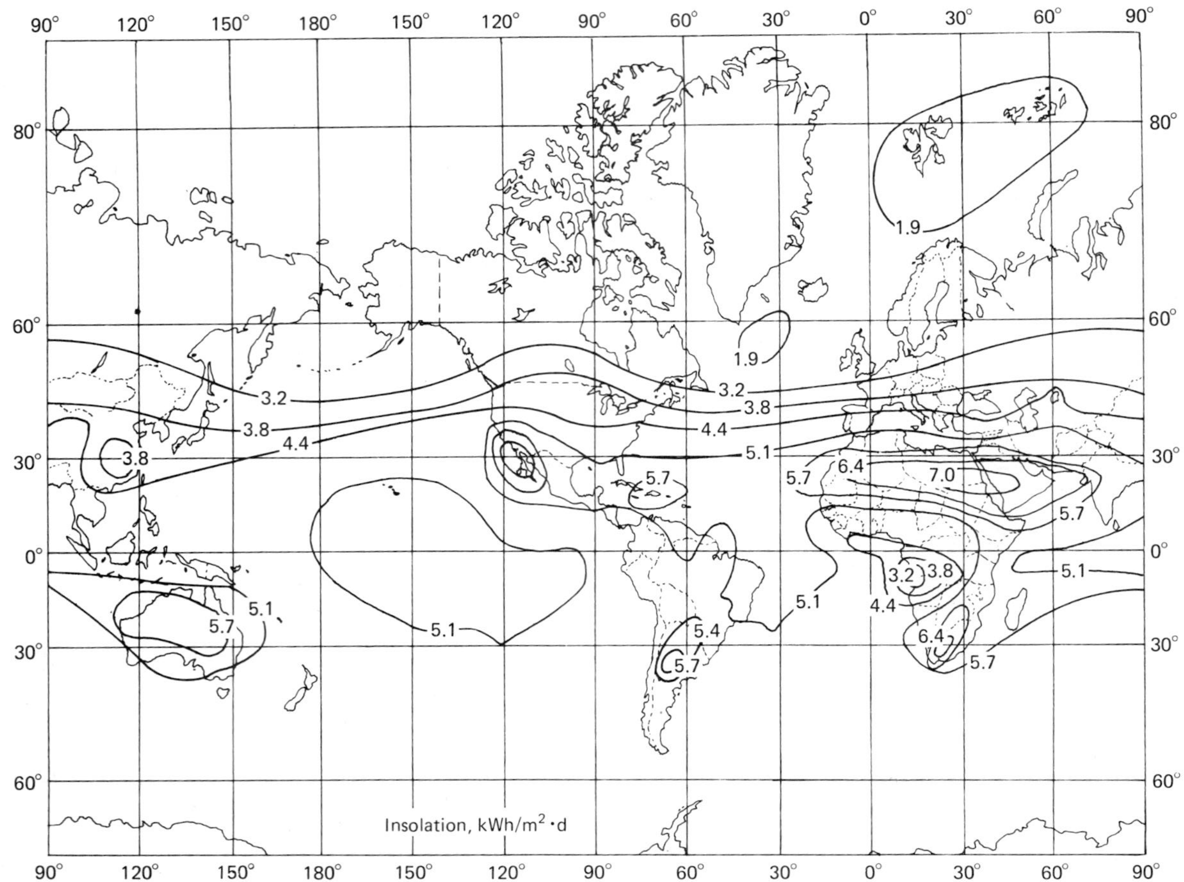

Figure 3-11 World annual average insolation on a horizontal surface, in kWh/m^2·day.

$$\delta = 23.45 \sin \left(360 \frac{284 + n}{365} \right) \tag{3-3}$$

$$\omega_s = \cos^{-1}(-\tan \phi \tan \delta) \tag{3-6}$$

$$\omega_s' = \min \begin{bmatrix} \cos^{-1}(-\tan \phi \tan \delta), \\ \cos^{-1}(-\tan (\phi - \beta) \tan \delta) \end{bmatrix} \tag{3-7}$$

$$\overline{R}_b = \frac{\cos (\phi - \beta) \cos \delta \sin \omega_s' + (\pi/180)\omega' \sin (\phi - \beta) \sin \delta}{\cos \phi \cos \delta \sin \omega_s + (\pi/180)\omega_s \sin \phi \sin \delta} \tag{3-8}$$

$$\frac{\overline{H}_D}{\overline{H}} = 1.390 - 4.027\overline{K}_T + 5.531\overline{K}_T^2 - 3.108\overline{K}_T^3 \tag{3-9}$$

$$\overline{R} = \frac{\overline{H}_T}{H} = \left(1 - \frac{\overline{H}_d}{H}\right)\overline{R}_b + \frac{\overline{H}_d}{H}\left(\frac{1 + \cos \beta}{2}\right) + \rho\left(\frac{1 - \cos \beta}{2}\right) \tag{3-10}$$

where δ = the sun's declination angle ($-23.45°$ to $+23.45°$)

n = day of year (1 to 365)

ω_s = sunset hour angle on a horizontal surface

ω_s' = sunset hour angle on a tilted surface

ϕ = the site's latitude angle (0° to 90°) in the Northern Hemisphere

β = tilt or slope angle of a south-facing surface (0 to 180°; $\beta > 90$ means that the surface is facing downward)

$\overline{R}_b$ = ratio of monthly average daily direct-beam radiation on a tilted south-facing surface to that on a horizontal surface

$\overline{H}_d$ = monthly average daily diffuse radiation on a horizontal surface

$\overline{H}$ = monthly average daily total radiation on a horizontal surface

$\overline{K}_t$ = clarity coefficient: the ratio of insolation on the earth to insolation directly outside the earth's atmosphere

$\overline{R}$ = ratio of monthly average daily total radiation on a tilted south-facing surface to that on a horizontal surface

$\overline{H}_T$ = monthly average daily total radiation on a tilted south-facing surface

ρ = reflection coefficient: the fraction of light reflected by a surface (0 to 1)

Daily Solar Power

Solar power received on earth varies throughout the day as a result of changes in the incident angle of sunlight on a surface, the amount of obstructing atmosphere sunlight must pierce, and the length of a day. A typical plot of solar power versus time on a clear day is shaped like a half sine wave with its crest at solar noon, as shown in Figure 3-12. On a partly cloudy day, the solar power level jumps up and down as the sun passes in and out behind clouds, so that the power/time-of-day plot becomes erratic, as shown in Figure 3-13. Notice in Figure 3-13 how solar power momentarily can rise above the normal clear-day power level during instances of natural concentration when indirect light is received from reflection off clouds.

The curve of solar power versus time of day varies for different days of the year. The number of sunlight hours and the clear-day peak power on a horizontal surface both reach their maximum on the summer solstice (June 21st) and drop to their minimum on the winter solstice (December 21st), as shown in Figure 3-12. The clear-day peak power on a latitude-tilted, south-facing surface reaches its maximum on the equi-

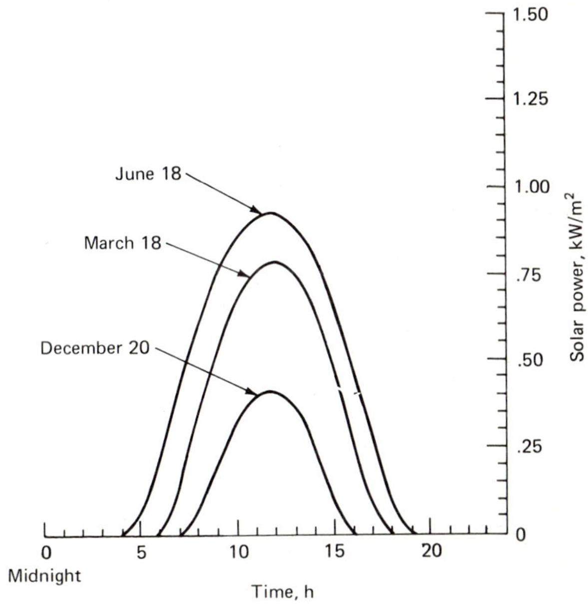

Figure 3-12 **Clear-day solar power on a horizontal surface versus time of day for three different days of the year. (M.I.T. Lincoln Laboratory, Concord, Mass., weather station.)**

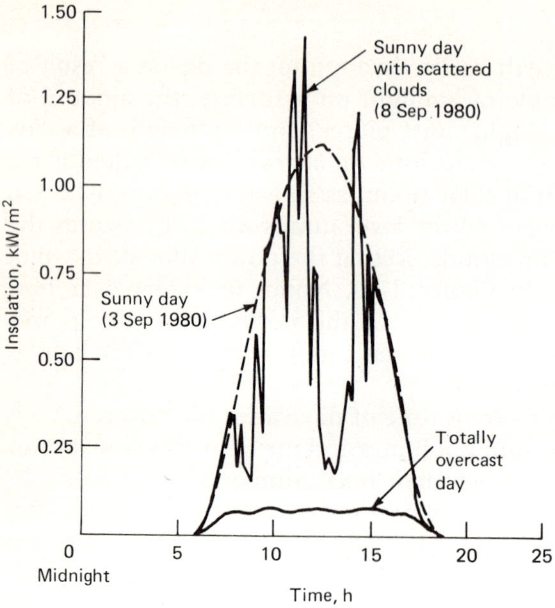

Figure 3-13 **Solar power on a latitude-tilted south-facing surface versus time of day for partly cloudy weather with natural sunlight concentration resulting from light being reflected off clouds. (M.I.T. Lincoln Laboratory, Natural Bridges National Monument, Utah, weather station.)**

noxes (March 21st and September 21st), as portrayed in Figure 3-14, because the sun's rays are most perpendicular to the tilted surface at that time of year (as previously demonstrated in Figure 3-7).

A solar-power/time-of-day curve can be estimated from three pieces of information: the sunrise time, the sunset time, and the peak power at solar noon. An approximate plot of solar power versus time of day will be useful when designing a photovoltaic array to satisfy a load demand.

Knowing the time of sunset and the length of day for a particular site requires knowing the site's latitude and the sun's declination. The nomogram in Figure 3-15, devised by Whillier, is a useful and simple tool for determining a day's length and sunset time, and is valid for sites in both the Northern and Southern Hemispheres. To use this chart, note the latitude and declination points for the place and time in question as they are found on their respective axes of the nomogram, then draw a line through these two points until the line touches the sunset hour axis.

The sunset time and length of day are read off the two charted lines as shown in the example. The sunrise time simply equals the sunset time minus the length of day. The length of day and time of sunrise and sunset can also be found in most newspapers and almanacs.

The average clear-day peak solar power at sea level on a south-facing surface tilted within 10° of the latitude angle is about 950 W/m² for latitudes between 0° and 50°. The sun's higher and lower path in the sky during different months and seasons of the year causes this value to fluctuate somewhat. Changes in altitude above sea level result in an increase in average clear-day peak power of about 20 W/m² per 1000 ft, or about 300 meters, of elevation, because at higher altitudes there is less atmosphere intercepting the sun's rays. Therefore, at an altitude of 5000 ft, or 150° m, the average clear-day peak power will be about 1050 W/m².

The sky is not always clear, for the sun's direct-beam rays often are blocked by clouds. When the sky is cloudy, the amount of sunlight that is readily convertible into electricity by a photovoltaic cell is substantially

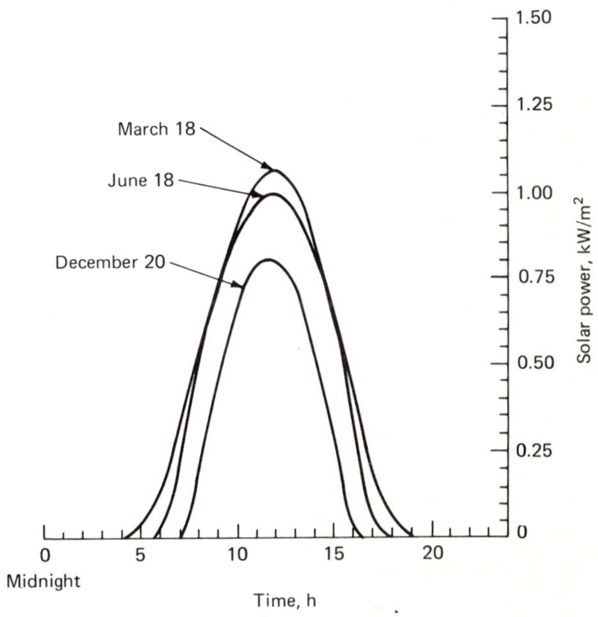

Figure 3-14 Clear-day solar power on a latitude-tilted south-facing surface versus time of day for three different days of the year (M.I.T. Lincoln Laboratory, Concord, Mass., weather station.)

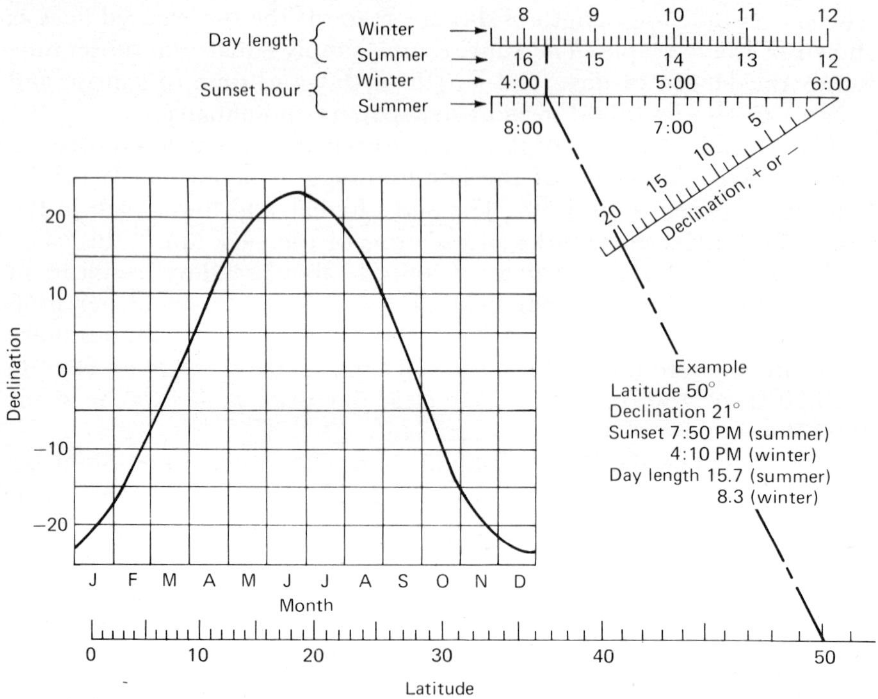

Figure 3-15 Whillier's nomogram for determining the length of day and time of sunset. (Adapted from: John A. Duffie and William A. Beckman, *Solar Engineering of Thermal Processes*, John Wiley & Sons, New York, 1980, p. 14; originally from Whillier, *Solar Energy* 9, Pergamon Press, Elmsford, N.Y., 1965, p. 164).

reduced, by at least 50% during thin cirrus cloud cover and by about 75% or more during denser cumulus cloud cover. The total sunlight hours divided by the total possible sunshine time is called the *percentage of possible sunshine*. This percentage, measured for different parts of the United States and different times of year, is given in Table 3-2. Multiplying the clear-day peak power for a particular time of year by the percentage of possible sunshine for that month will yield an approximate average peak power for the month that will be sufficiently accurate for our purposes. This estimate will overlook the effects of airborne dust, pollution, or humidity; if any of these conditions becomes excessive, it must be measured and adequately accounted for.

We have three data points on a solar-power/time-of-day curve that can be plotted on a graph. At sunrise and sunset the solar power is zero, while at solar noon the solar power is at its peak, as estimated using the technique outlined above. If these three data points are plotted on a

Table 3-2 Average Percentage of Possible Sunshine: Selected Cities.*

State	Station	Length of record, years	Jan	Feb	Mar	Apr	May	June	July	Aug	Sept	Oct	Nov	Dec	Annual
AL	Montgomery	25	47	53	58	64	66	65	63	65	63	66	57	50	59
AK	Juneau	30	33	32	37	38	38	34	30	30	25	19	23	20	31
AZ	Phoenix	80	78	80	83	89	93	94	85	85	89	88	83	77	86
AR	Little Rock	32	46	54	57	61	68	73	71	73	68	69	56	48	63
CA	Los Angeles	32	69	72	73	70	66	65	82	83	79	73	74	71	73
	Sacramento	27	45	61	70	80	86	92	97	96	94	84	64	46	79
	San Francisco	38	56	62	69	73	72	73	66	65	72	70	62	53	67
CO	Denver	26	72	71	69	66	64	70	70	72	75	73	65	68	70
CN	Hartford	21	58	57	56	57	58	58	61	63	59	58	46	48	57
DE	Wilmington†	25	50	54	57	57	59	64	63	61	60	60	54	51	53
DC	Washington	27	48	51	55	56	58	64	62	62	62	60	53	47	57
FL	Jacksonville	25	57	61	66	71	69	61	59	58	53	56	61	56	61
	Key West	17	72	76	81	84	80	71	75	76	69	68	72	74	75
GA	Atlanta	41	47	52	57	65	69	67	61	65	63	67	60	50	61
HI	Honolulu	23	63	65	69	67	70	71	74	75	75	67	60	59	68
ID	Boise	35	41	52	63	68	71	75	89	85	82	67	45	39	67
IL	Chicago	33	44	47	51	53	61	67	70	68	63	62	41	38	57
	Peoria	32	45	50	52	55	59	66	68	67	64	63	44	39	57
IN	Indianapolis	32	41	51	51	55	61	68	70	71	66	64	42	39	58
IO	Des Moines	25	51	54	54	55	60	67	71	70	64	64	49	45	59
KS	Wichita	22	59	59	60	62	64	69	74	73	65	66	59	56	65
KY	Louisville	28	41	47	50	55	62	67	66	68	65	63	47	39	57
LA	Shreveport	23	49	54	56	55	64	71	74	72	68	71	62	53	64
ME	Portland	35	55	59	56	56	56	60	64	65	61	58	47	53	58
MD	Baltimore	25	51	55	55	55	57	62	65	62	60	59	51	48	57

Table 3-2 (continued)

State	Station	Length of record, years	Jan	Feb	Mar	Apr	May	June	July	Aug	Sept	Oct	Nov	Dec	Annual
MA	Boston	40	54	56	57	56	58	63	66	67	63	61	51	52	59
MI	Detroit	32	32	43	49	52	59	65	70	65	61	56	35	32	54
	Sault St. Marie	34	34	46	55	55	56	57	63	58	46	41	23	28	48
MN	Duluth	25	49	54	56	54	55	58	67	61	52	48	34	39	54
	Minneapolis-St. Paul	37	51	57	54	55	58	63	70	67	61	57	39	40	58
MS	Jackson	11	48	55	61	60	63	67	61	62	58	65	54	45	59
MO	Kansas City	3	64	54	61	65	67	72	84	69	51	62	46	54	64
	St. Louis	16	52	51	54	56	62	69	71	66	63	62	49	41	58
MT	Great Falls	33	49	57	67	62	64	65	81	78	68	61	46	46	64
NE	Omaha	40	55	55	55	59	62	68	76	72	67	67	52	48	62
NV	Reno	33	66	68	74	80	81	85	92	93	92	83	70	63	80
NH	Concord	34	52	54	52	53	54	57	62	60	54	54	42	47	54
NJ	Atlantic City	15	49	48	51	53	54	58	60	62	59	57	50	42	54
NM	Albuquerque	36	73	73	74	77	80	83	76	76	80	79	78	72	77
NY	Albany	37	46	51	52	53	55	59	64	61	56	53	36	38	53
	Buffalo	32	34	40	46	52	58	66	69	66	60	53	29	27	53
	New York‡	99	50	55	56	59	61	64	65	64	63	61	52	49	59
NC	Charlotte	25	55	59	63	70	69	71	68	70	68	69	63	58	66
	Raleigh	21	55	58	63	64	60	61	61	61	60	63	63	56	60
ND	Bismarck	36	54	56	60	58	63	64	76	73	65	59	44	47	62
OH	Cincinnati	60	41	45	51	55	61	67	68	67	66	59	44	38	57
	Cleveland	34	32	37	44	53	59	65	68	64	60	55	31	26	52
	Columbus	24	37	41	44	52	58	62	64	63	62	58	38	30	53

OK	Oklahoma City	23	59	61	63	63	65	73	75	77	69	68	60	59	67
OR	Portland	26	24	35	42	48	54	51	69	64	60	40	27	20	47
PA	Philadelphia	33	50	53	56	56	57	63	63	63	60	60	53	49	58
	Pittsburgh	23	36	38	45	48	53	60	62	60	60	56	40	30	50
RI	Providence	22	57	56	55	55	57	57	59	59	58	60	49	51	56
SC	Columbia	22	56	59	64	67	66	65	64	65	65	66	64	60	63
SD	Rapid City	33	54	59	61	59	57	60	71	73	67	65	56	54	62
TN	Memphis	25	48	54	57	63	69	73	72	75	69	71	58	49	64
	Nashville	33	40	47	52	59	62	67	64	66	63	64	50	40	57
TX	Amarillo	34	69	68	71	73	73	77	77	78	74	75	73	67	73
	El Paso	33	78	82	85	87	89	89	79	80	82	84	83	78	83
	Houston	6	41	54	48	51	57	63	68	61	57	61	58	69	56
UT	Salt Lake City	38	47	55	64	66	73	78	84	83	84	73	54	44	70
VT	Burlington	32	42	48	52	50	56	60	65	62	55	50	30	33	51
VA	Norfolk	19	57	58	63	66	67	68	65	65	64	60	60	57	63
	Richmond	25	51	54	59	62	64	67	65	64	63	59	56	51	60
WA	Seattle-Tacoma	10	21	42	49	51	58	54	67	65	61	42	27	17	49
	Spokane	27	26	41	53	60	63	65	81	78	71	51	28	20	57
WV	Parkersburg	78	32	36	43	49	56	59	62	60	59	54	37	29	48
WI	Milwaukee	35	44	47	51	54	59	63	70	67	60	57	41	38	56
WY	Cheyenne	40	61	65	64	61	59	65	68	68	69	68	60	59	64
PR	San Juan	20	65	69	74	69	61	57	64	65	59	59	57	56	63

Source: U.S. National Oceanic and Atmospheric Administration; reproduced in *Solar Energy Handbook: Theory and Application* by Power Systems Group, Ametek, Inc., published by Chilton Book Co., 1979, p. 46.

*Average data, except as noted, for period of record through 1975.

†Data not available; figures are for a nearby station.

‡City office data.

suitably scaled graph, they can be interconnected by drawing a curve that approximates the shape of the clear-day power curves shown in Figure 3-12. Our plot is only a monthly average and does not trace the sudden changes in power level that occur on partly cloudy days. This technique assumes that cloud cover is spread out fairly uniformly throughout the day, which often is not the case.

4

THE SILICON SOLAR CELL RESPONSE

Having a silicon solar cell before us for the first time, naturally we are puzzled about how it operates. We know that it has two polarities much like a battery, one positive and the other negative. Also, we have been informed that electricity is produced when the top of the cell is exposed to light. Nonetheless, many questions remain. How much energy can we expect to obtain from a photovoltaic cell? Under what conditions is it most productive?

A SOLAR-CELL EXPERIMENT

A good way to discover how a solar cell performs is to subject it to special tests. Terrestrial solar cells experience various changing conditions, some of which increase their output and others of which limit it. There are four variables that have a significant effect: light, temperature, cell area, and load. Light strikes a cell in varying quantities, a solar cell is exposed to a wide range of temperatures, solar cells are manufactured with various dimensions or amounts of surface area, and power-consuming devices or loads of various sizes are attached to the terminals of this solar-power generator. Our knowledge of photovoltaics would be advanced greatly if we could isolate these individual conditions and let them affect our cell one at a time. A scientifically sound procedure would be to hold

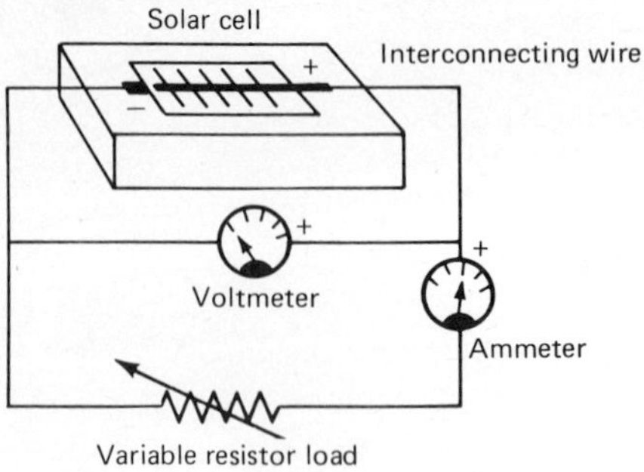

Figure 4-1 **The solar-cell test setup.**

all but one of the changing states steady while manipulating the remaining variable.

To save time and to increase the accuracy of our experiment, it would be convenient if we could move the sun about in the sky at will or at times keep its position fixed. We want to vary the temperature over a wide range, so it would be handy if we could switch from summer to winter and bring in chilling frosts or stifling heat. We will have cells of various sizes at hand to experiment with, and as a load we will use a variable resistor. The test site will be outdoors at sea level in a wide-open area with no obstructions or shade. Our solar cell is laid down flat on a level platform.

We need some instruments to measure the performance of our solar cell. A voltage meter (voltmeter) is attached across the terminals of the solar cell, and a current meter (ammeter) is placed in series with the load. Power output can be calculated by multiplying the voltage and current readings together.

The test setup is complete as shown in Figure 4-1. We will conduct the experiment carefully, concentrating on one variable at a time.

The Light Variable

At solar noon on a clear day, a site on earth experiences that day's maximum solar power, because direct solar radiation has the least amount of atmosphere to pierce and its incident angle with the earth's surface is closest to the perpendicular. As the sun's position moves away from its solar-noon apex, its intensity decreases. Therefore, by moving the sun around the sky, we are in effect varying the intensity of the light striking our solar cell.

Let us set the climate to a spring day, sweeping away any clouds or haze to produce a clear sky. The temperature is comfortable at 20°C.

The movable sun is our variable light source. We will work with only one cell, so the cell surface area will remain constant. Our load will have only two possible states: open circuit or infinite resistance, and short circuit or zero resistance. Finally, we shall begin our first test.

Moving the sun below the horizon, we notice that both the voltmeter and the ammeter read zero. As the sun begins to rise, the solar-cell voltage quickly jumps to about 1/2 volt. Moving the sun higher in the sky and thus increasing its intensity, we find that the voltage increases only a small amount, reaching a peak of around 6/10 volt. The ammeter reads no current, since the load is disconnected and there is no path for electron flow. From this test we can conclude that *a solar cell's voltage varies little with light intensity, except in the beginning when light is initially applied.* A graph of insolation versus voltage is shown in Figure 4-2.

Now let us put a short circuit or zero resistance across our cell. With no light or the sun out of sight, both voltmeter and ammeter read zero. As the sun rises into the sky, we notice the current increasing at a similar rate. In fact, if we double the light intensity, the current also doubles! A basic rule is established: *Solar-cell current is directly proportional to light intensity.* This fact has made photovoltaics useful as a light-measuring device in such instruments as a camera light meter. The voltmeter reads zero when the solar cell is shorted out because it is not possible to measure a voltage across zero resistance. A graph of current versus insolation is also shown in Figure 4-2.

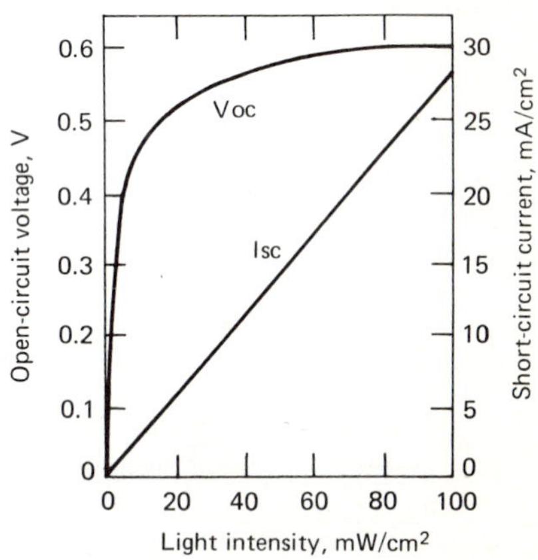

Figure 4-2 **Voltage and current versus light intensity for a silicon solar cell. (From: Joseph A. Merrigan, *Sunlight to Electricity*, © M.I.T. Press, Cambridge, Mass., 1975, p. 52.)**

The Temperature Variable

In this test we will vary the earth's climate. Winter will be exchanged for summer for the purpose of temperature-cycling our solar cell. Throughout this harsh contrast, the sun will remain fixed high in the sky. We will continue to work with the same cell surface area, and the load will be switched between open and short circuit.

Beginning with an open circuit at 20°C, we notice that the voltmeter reads 6/10 volt and the ammeter is resting undisturbed at zero. Fluctuating the climate between winter and summer can produce changes in temperature of as much as 60°C. How does our photovoltaic cell suffer under these conditions? Heat is a form of energy, so should not our cell become more energetic as the temperature rises? On the contrary, the silicon-cell voltage drops at an approximate rate of 2.2 mV for every 1°C increase. Adjusting the temperature to a low level, we notice the cell voltage climbing. This test demonstrates that *cell voltage is inversely proportional to temperature,* or, in other words, *a rise in temperature produces a proportional decrease in voltage.* A plot of voltage versus temperature for silicon solar cells is found in Figure 4-3.

Now let us put a short circuit across our cell and learn how temperature affects cell current. The voltmeter has dropped to zero, while the ammeter hovers around 1 ampere. Moving back and forth from a winter frost to a summer heat wave, we notice little movement of the ammeter needle. *Solar-cell current is relatively steady despite changing temperatures.* Since power is the product of current and voltage, we can deduce a simple fact: *Solar-cell power output increases as cell temperature decreases.*

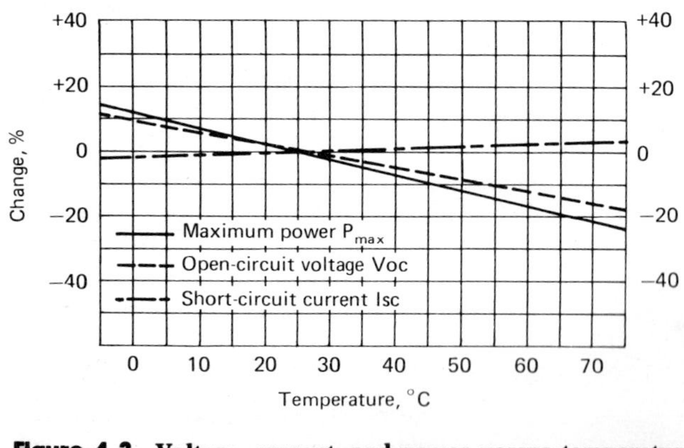

Figure 4-3 **Voltage, current, and power versus temperature for a silicon solar cell. (From: Technical Staff of Solarex Corporation, *Making and Using Electricity from the Sun,* Tab Books, Blue Ridge Summit, Pa., 1979, pp. 39, 41.)**

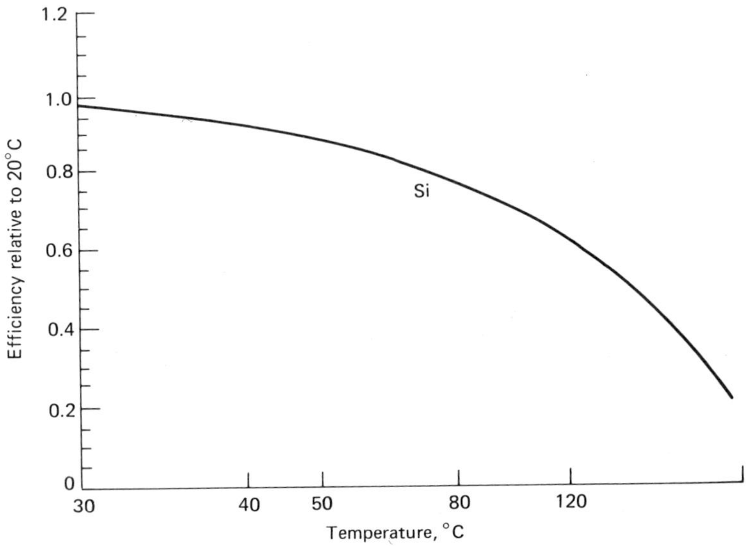

Figure 4-4 Silicon solar-cell efficiency versus temperature. (From: Woodall and Hovel, *Journal of Vacuum Science & Technology*, 12, American Institute of Physics, N.Y., 1975, p. 1000.)

Figure 4-3 is a graph of silicon-cell and power versus temperature, while Figure 4-4 shows how the conversion efficiency of a silicon cell is altered by temperature.

To design a photovoltaic system, we will need to predict the solar-cell voltage and power output under different circumstances. As we have just learned, solar-cell voltage and power are significantly influenced by cell temperature. At this point it would be valuable to examine how cell temperature varies under different environmental conditions.

The daytime temperature of a solar cell is not simply equal to the ambient temperature, since solar cells are dark in color and therefore absorb a greater portion of the sun's energy. During the day, a solar cell operates hotter than the ambient temperature by a factor that depends on insolation and on the cell's packaging and mounting. If a solar cell is arranged in a way that facilitates convective cooling, it will operate at a lower temperature. For this reason, the amount of wind at a site can also affect cell temperature (the effect of wind will, however, be ignored because it is small under average wind conditions and because daytime wind information is generally not available). A solar cell will operate slightly cooler when it is delivering electricity than when it is not. As will be discussed later, when a suitably sized load is attached to an illuminated solar cell, it will deliver 5% to 15% of the sun's energy to the load in the form of electricity. When this conversion is occurring, the solar cell will commensurately operate about 3°C cooler at one sun, since some of the sun's energy that would normally heat the cell has been transfered to the load.

Solar-cell temperature varies quite linearly with changes in insolation, as shown in Equation (4-1) and the plot in Figure 4-5.

$$T_c = T_a + kS \tag{4-1}$$

where T_c = cell temperature under no wind or load conditions, °C

T_a = ambient temperature, °C

k = solar-cell temperature coefficient, $\Delta T_c/\Delta S$

S = solar insolation, mW/cm²

The array temperature coefficient k indicates how much the cell temperature rises for a given increase in insolation; it varies from one solar cell and its mounting scheme to the next, being larger when there is poor ventilation and vice versa. Cell and ambient temperature and insolation measurements must be taken for each application, and a cell minus ambient temperature versus insolation plot, as shown in Figure 4-5, must be made in order to accurately determine the specific temperature coefficient. The k value is equal to the slope $[\Delta(T_c - T_a)/\Delta S]$ of the aforementioned plot and typically is between 0.2 and 0.4.

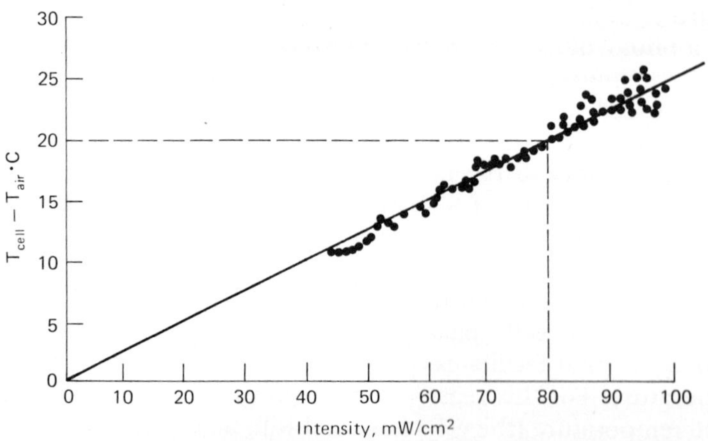

Figure 4-5 **Solar-cell minus ambient temperature versus insolation. (Low Cost Solar Array Project, *Block IV Solar Cell Module Design and Test Specification for Residential Application*, Jet Propulsion Laboratory, Pasadena, Calif., 1978, p. A3.)**

A standard cell temperature called the *nominal operating cell temperature,* or NOCT, is generally used to rate solar-cell efficiency under somewhat realistic conditions. Manufacturers are often disposed to rate the peak power output and cell efficiency of their modules at low temperatures, between 15° and 30°C, because performance is better at these levels. This practice can be deceptive; the designer must always qualify efficiency and peak power ratings by the cell temperature at which they are measured, and usually should normalize the specifications to NOCT. The nominal operating cell temperature is measured under the following conditions: A clear-day average insolation of 80 mW/cm^2, an ambient temperature of 20°C, an average wind speed of 1 m/s, and open-circuit conditions. (*Note:* NOCT is measured at open circuit, while the cell temperature decreases about 3°C at one sun when power is being delivered to a load.) On a clear sunny day, about 50% of the solar energy is available above and below the 80 mW/cm^2 power level; by measuring NOCT at this insolation, we are in effect determining the cell temperature at the mean solar-power level. The nominal operating cell temperature will vary depending on how solar cells are packaged in a module, how modules are mounted under normal operating conditions, and how dark the cells are. Typical values of NOCT are between 40° and 70°C, being on the low end of the scale when the module design and support structure allow for easy convective cooling and vice versa.

Having examined the various considerations that influence cell temperature, we are prepared to address the more vital issues of changes in cell voltage and power. If photovoltaic power and voltage are rated at an unsuitable temperature, an adjustment will have to be made. Equations (4-2) and (4-3) can be used to perform these necessary adjustments.

$$V_a = V_{ref} \times [1 + \beta (T_{ref} - T_a)] \tag{4-2}$$

where V_a = adjusted solar-cell voltage, V

V_{ref} = reference solar-cell voltage, V

β = solar-cell output versus temperature coefficient, $V/V{\cdot}°C$

T_{ref} = reference cell temperature, °C

T_a = adjusted cell temperature, °C

$$P_a = P_{ref} \times [1 + \beta (T_{ref} - T_a)] \tag{4-3}$$

where P_a = adjusted solar-cell power, Watts

P_{ref} = reference solar-cell power, Watts

Solar-cell power and voltage vary with changes in temperature in almost the same manner, and therefore can adequately be served by the same coefficient, β. The solar-cell output/temperature coefficient β is unitless and varies somewhat from one cell to the next, but normally is between 0.004 and 0.006 per degree centigrade. This coefficient can be experimentally determined by measuring the fraction change in the open-circuit voltage of a solar cell as a result of actual changes in cell temperature within the relatively linear −20° to +80°C range. (As will be explained later, the open-circuit and maximum-power voltage of a solar cell vary in almost an identical way; since the open-circuit voltage is easier to measure, it will be the most suitable gauge.)

The Cell Area Variable

In this case, we choose to lock the sun into a position directly over the cell and to orchestrate a clear sky with a temperature of 20°C. Let us take a cell with an area of 1 cm² and install it on our platform. At open circuit the voltmeter reads 6/10 volt, and at short circuit the ammeter indicates 16/1000 ampere. In our collection of solar cells, we find one with an area of 10 cm². The first cell is replaced by one that is 10 times its size. At open circuit, the voltmeter still reads 6/10 volt. Evidently *cell area has no effect on cell voltage.* Putting a short circuit across the solar cell, we notice a dramatic change. The ammeter now reads 160/1000 ampere, a tenfold jump from our previous test. Our experiment has shown that *cell current is proportional to cell area*, and that in fact a tenfold increase in cell area results in the same growth in cell current. Multiplying the current and voltage together, we notice that *solar-cell power is similarly proportional to area.*

We have accumulated enough information to perform another vital test on our cell. How efficiently does our 1-cm² solar cell convert sunlight into electricity? We know the sun is poised such that 1,000 W/m² or 0.1 W/cm² of solar energy is striking the cell. Until now, we have not attached a load to extract power from our cell; we will learn shortly that the load's size will determine how much energy is actually delivered by the solar cell. If the 1-cm² cell delivers 0.01 W to an optimally sized load at a cell temperature of 40°C, its efficiency at that temperature will be 0.01/0.1 × 100 or 10%. When discussing efficiency, we must be aware of the related cell temperature and will assume the load was optimally sized. The determination of what load size is optimal is an important issue which must now be explored.

The Load Variable

The solar cell was developed for the sole purpose of supplying people with energy in a form that easily can do work for them. Since satisfying a load is the solar cell's reason for being, the way it responds to loads is a most vital piece of information indeed. We will approach this subject slowly, and will build on our knowledge as the book progresses.

Up until now we have allowed ourselves only two options for loading our cell: open circuit and short circuit. Let us begin to explore what exists between these two points. We will maintain the sun directly overhead and establish the climate as clear, dry, and warm. The 10-cm² cell is mounted on the test block and wired as shown in Figure 4-1.

Our experiment begins with an open circuit across the solar-cell terminals. As before, the voltmeter reads 6/10 volt and the ammeter shows nothing. As we slowly vary the resistor load from infinite to zero resistance, we observe the cell current climbing and the cell voltage dropping. It is hard to imagine what is going on here. To get a better picture, we will plot the solar-cell response on a graph. The curve that results from plotting current against voltage as the load varies from open to short circuit is shown in Figure 4-6. Examine this curve well, for it will reappear.

A solar cell acts like a constant current source for most of its curve, in that it provides a constant current and varies the voltage enough to maintain this particular current despite a changing load. As the resistance continues to increase, a breakdown point is reached at which the cell can no longer maintain the established level of current, and the curve drops off to zero. The curve displayed in Figure 4-6 is a current-versus-voltage curve. Since current is usually represented by the symıool

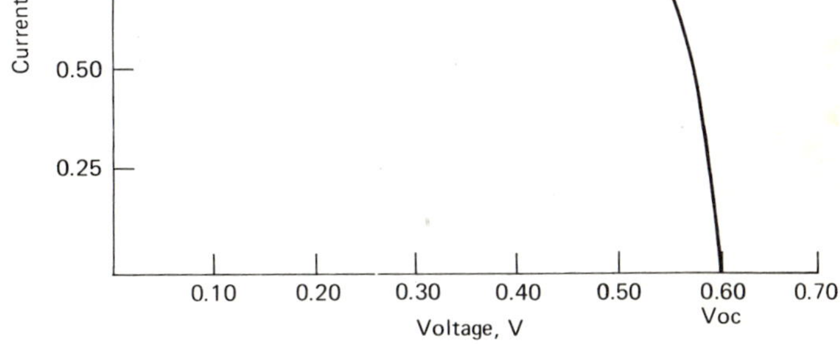

Figure 4-6 **A silicon solar-cell current-versus-voltage curve.**

I and voltage by the symbol *V*, we commonly call it an *I/V curve.* The open-circuit voltage *Voc* is that point at which the curve touches the *V* axis, and the short-circuit current *Isc* is that point at which the curve touches the *I* axis.

Measuring current and voltage is fine, but what most of us are interested in is power. Much of the world's work is done with electric power, and it is this that we are hoping to get from our solar cell. Since power is the product of voltage and current, this information is readily at hand.

Let us run our illuminated solar cell through another sweep of the variable load, but this time we will plot power against voltage. At open circuit and short circuit we will have no power, since in the first case the current is zero and in the latter case the voltage is shorted out. Starting at open circuit and varying the resistor toward a short circuit, we take a series of data points. As we calculate power and plot it against voltage, the curve begins to unfold. The power curve, as shown in Figure 4-7, is shaped like a mountain with one peak. At the top of the mountain, we are getting the most power possible. On either side, it is downhill all the way. Evidently, there is only one value of resistance that gleans the most power from our cell. Any other load will obtain less than the maximum.

Since good sense would dictate that we set up our solar cell to get the most out of it, we want to learn more about the maximum-power point. In order to find the voltage, current, and resistance values for this precious point, we are going to merge two graphs. Let us lay the power curve on top of the *I/V* curve for the same cell so that the voltage axes

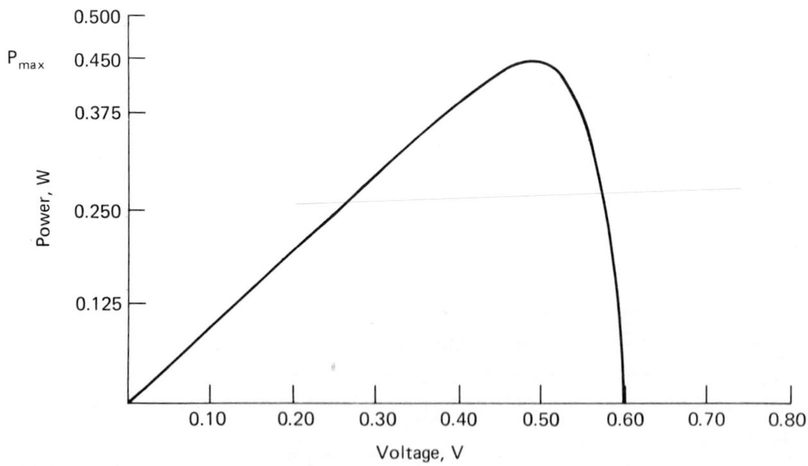

Figure 4-7 A silicon solar-cell power-versus-voltage curve.

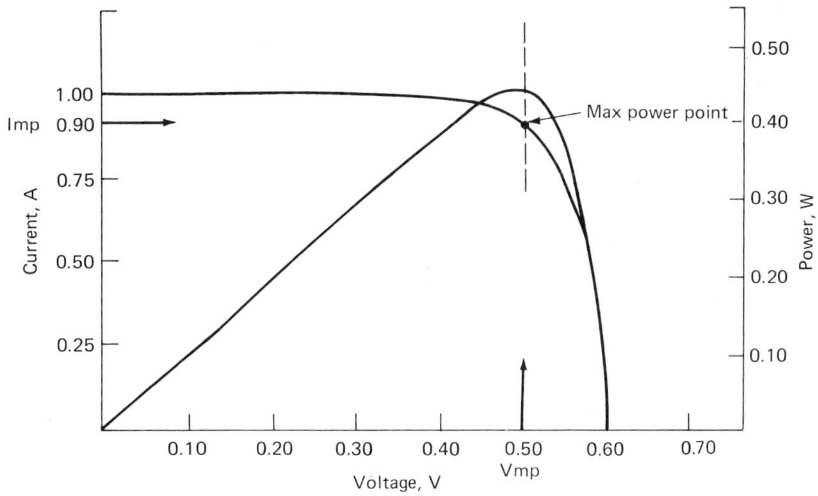

Figure 4-8 Finding the maximum-power current and voltage for a silicon solar cell at a given insolation.

of the two graphs line up perfectly. Then we draw a faint perpendicular line from the crest of our power curve to where it crosses the *I/V* curve, as shown in Figure 4-8. From this point of intersection, we can read off the maximum power current and voltage (*Imp* and *Vmp*). We then obtain the resistance value for this maximum-power point by dividing the voltage by the current. By using this value of resistor, we are getting the optimum performance from the solar cell.

THE SOLAR-CELL RESPONSE UNDER NORMAL CONDITIONS

Since the sun and weather usually do not bow to our every wish and because the light, temperature, and load variables do not remain constant, we must relax our imaginary hold over them. Let us observe what happens to the *I/V* curve when normal daily and seasonal fluctuations occur. In the dark of night, our curve lies dormant on the graph's zero point. As the sun rises, so does our *I/V* curve. When dawn breaks, the voltage jumps to about 1/2 volt, and the horizontal arm of the curve, indicating cell current, climbs with increasing amounts of sunlight, as shown in Figure 4-9. The curve expands to its highest point around noon, then gradually sinks on its way to sunset. Changes in temperature throughout a day or from season to season affect the relatively vertical portion of the curve, as demonstrated in Figure 4-10. Increasing tem-

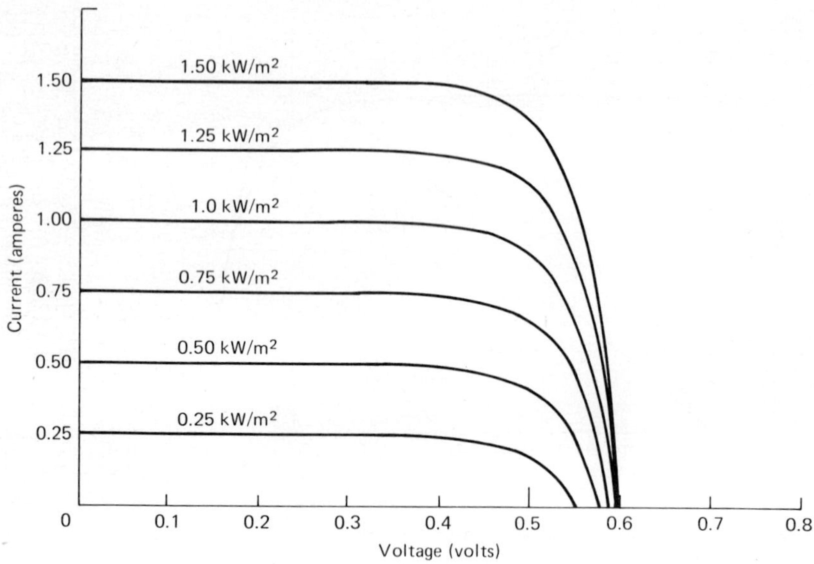

Figure 4-9 **Changing current with changing insolation.**

perature causes the voltage to drop and the vertical part of the *I/V* curve to move leftward, while decreasing temperature produces the opposite effect.

The *I/V* curve displays how a solar cell will respond to any and all possible loads under a particular set of light and temperature conditions.

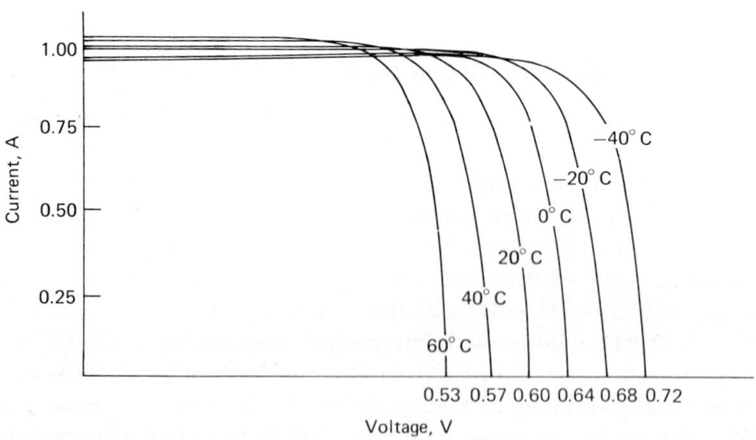

Figure 4-10 **Changing voltage with changing temperature.**

When a fixed resistive load is attached to a photovoltaic, it represents itself as a point somewhere on this curve. As our load increases or the resistance decreases, the operating point moves along the curve toward the left. An increase in resistance causes the operating point to ride the curve in a rightward direction, as shown in Figure 4-11.

It is the aim of all designers of solar cells to make the load operating point coincide with the maximum-power point. In a static world this match would not be difficult. Problems arise when we let nature go through its rhythm. The operating and maximum-power points may be aligned at noon, but will be separated in the morning and late afternoon. Look closely at how the operating and peak power points move as the sun's insolation changes. Notice in Figure 4-12 how the operating point on the rising or sinking *I/V* curve traces a straight line that ultimately has its origin at the graph's zero point. What is the meaning of this line? The line's slope (voltage divided by current) is exactly equal to the resistance value of our load. This line is aptly called the *load line* and is no more or less than a graphic representation of the resistance value we have across the photovoltaic. A solar cell's maximum-power point has an obstinately different response to changing insolation. While the operating point wanders off at an angle, the maximum-power point travels a relatively vertical path that leans slightly to the right, as displayed in Figure 4-12 (the slope of this path depends on the *I/V* curve fill factor, as will be explained later). Here we have a dilemma: The resistance load may be tuned to optimize the power received from our solar cell at

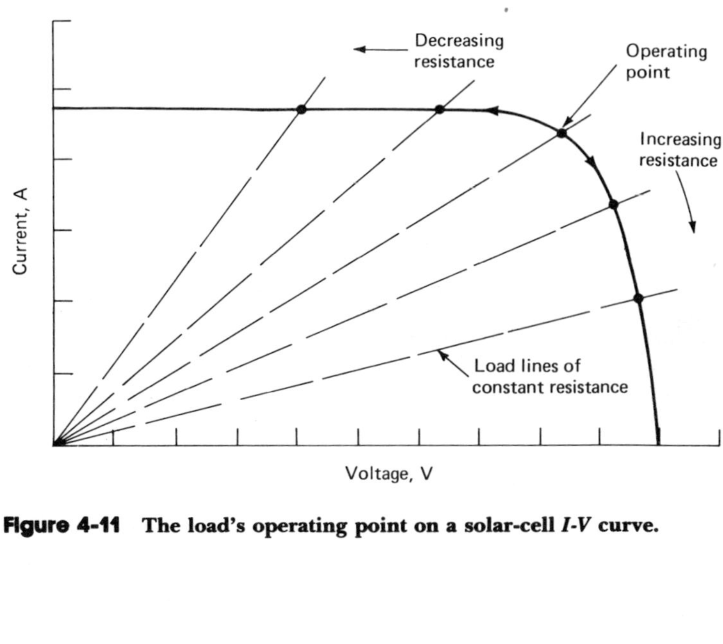

Figure 4-11 **The load's operating point on a solar-cell *I-V* curve.**

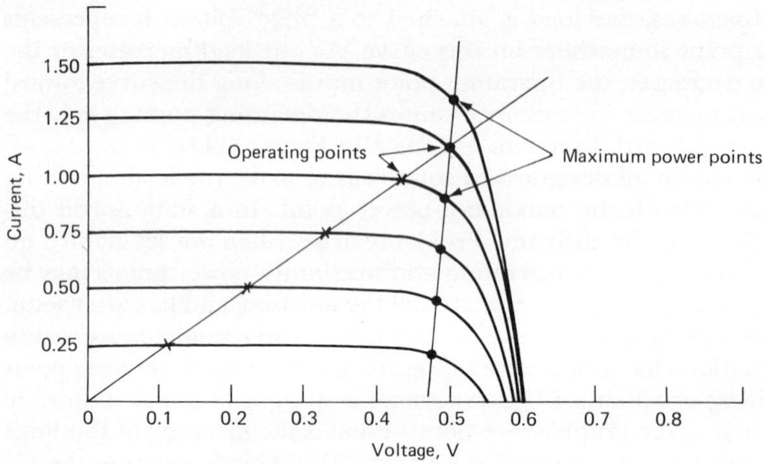

Figure 4-12 The resistive load's operating point and the PV cell's maximum-power point travel divergent paths as insolation changes.

breakfast, but be miserably out of tune by lunchtime. The riddle of how to bring the maximum-power and operating points closer together will be discussed in a later chapter that pertains to the photovoltaic system.

SOLAR-CELL SIMULATION

A common practice in science and engineering is to make an equivalent model of a device or system so as to better analyze and predict its performance. Let us develop an equivalent circuit for a solar cell, and provide a computer program that can simulate the cell's output characteristics.

As the *I/V* curve demonstrates, a solar cell acts like a constant current source under low-resistance load conditions (at any particular insolation). As the load resistance increases, the solar cell fails in its attempt to maintain a fixed current, and the current drops off to zero. We can start by representing our solar cell as an ideal current source.

The PN junction used to produce photovoltaic electricity is also used as a device that allows current flow only in one direction, called a *diode* (current in a diode is able to flow from the positive terminal to the negative with only a 6/10-volt drop, but is inhibited from flowing in the opposite direction). The only fundamental difference between a photovoltaic PN junction and a diode PN junction is that the former is active and produces energy, while the latter is passive and absorbs energy; their *I/V* curves are identical in shape but are located in different quadrants of a graph, as shown in Figure 4-13. A diode wired across an ideal current source as in Figure 4-15 exhibits a current-versus-voltage response curve that perfectly mimics a photovoltaic cell's *I*/*V* characteristics.

A typical solar cell has imperfections that can be represented by *series* and *shunt resistance.* The resistance that exists between the solar cell and its load is shown in our model as a series resistor, while the leakage resistance between the positive and negative terminals of a cell is represented as a parallel resistor. A solar-cell's series resistance should be as low as possible, while its shunt resistance should be very high. The two graphs in Figure 4-14 show how changes in series and shunt resistance affect the solar-cell *I*/*V* curve. The *fill factor* of a solar cell is defined as the ratio of peak power to the power calculated by multiplying the open-circuit voltage by the short-circuit currents and is a reflection of how much series resistance and how little shunt resistance there is in a solar cell and its circuit. A healthy fill factor for a silicon solar cell ranges between 0.6 and 0.8. As a solar cell degrades with age, its series resistance tends to increase, resulting in a lower fill factor. Fill factor affects how

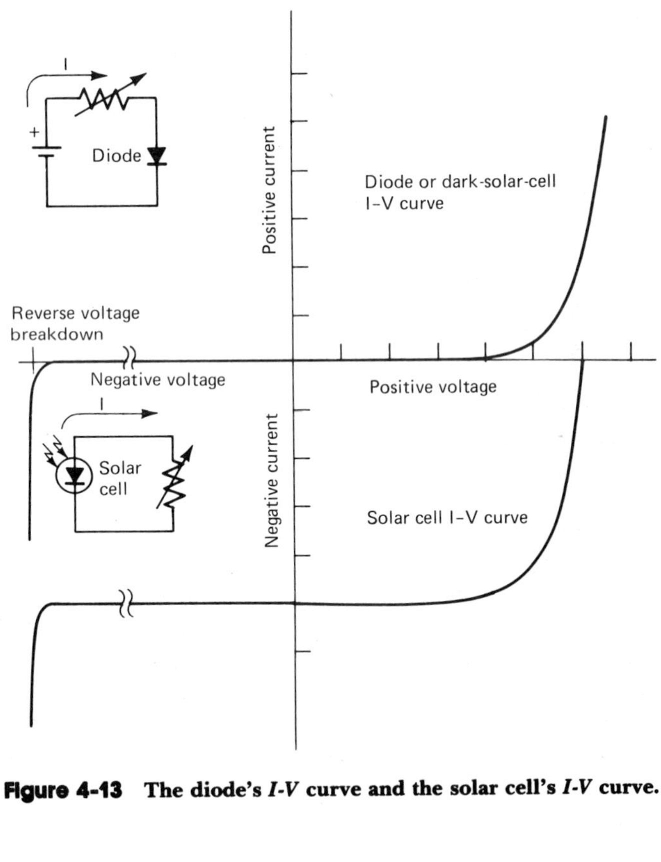

Figure 4-13 The diode's *I-V* curve and the solar cell's *I-V* curve.

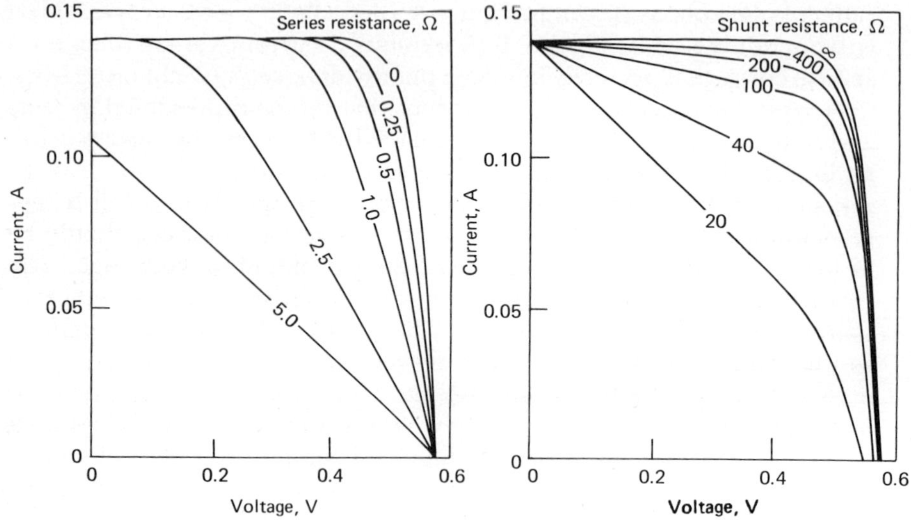

Figure 4-14 **The *I-V*-curve response to changes in series and shunt resistance. (From: H. S. Rauschenbach, *Solar Cell Array Design Handbook,* Van Nostrand Reinhold Company, New York, copyright © 1980, p. 170; reprinted by permission of the publisher.)**

a solar cell's maximum-power voltage changes with insolation. As shown in Figure 4-12, the maximum-power voltage remains relatively stable through changes in insolation with a normal cell that has a fill factor of about 0.7. As the fill factor decreases, however, this stability changes, and the relatively vertical path traveled by the maximum-power voltage takes a tilt toward the right; an increase in insolation produces a larger increase in the maximum-power voltage. This fact will become more relevant as we try to maximize the power output of a photovoltaic system.

The current source, diode, series resistor, and shunt resistor are connected together in a solar-cell equivalent circuit as shown in Figure 4-15. This analytical model will prove both helpful in understanding how a solar cell operates and useful in identifying the types of problems that might occur with a solar-cell array as we try to troubleshoot for defects or malfunctions in a later chapter.

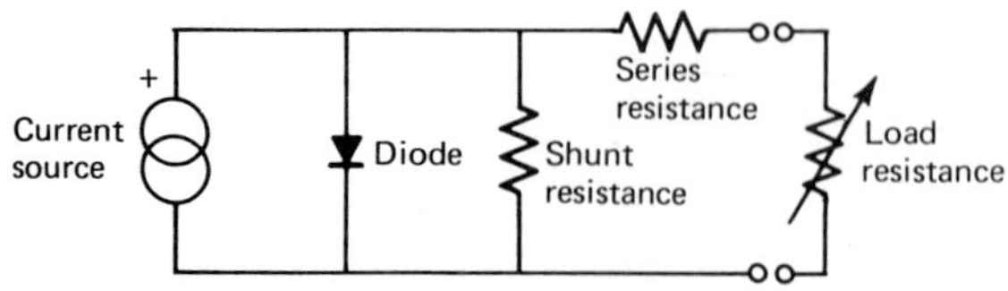

Figure 4-15 **A solar-cell equivalent circuit.**

Later design work will require simulating a photovoltaic cell's or array's output characteristics under different conditions of temperature and insolation. A computer program developed at M.I.T. using the Sandstrom and TRW model will serve this purpose. Seven factors must be known about a solar cell or array before we can consider plotting its *I/V* curve. A reference curve must initially be taken to determine the cell's or array's open-circuit voltage, short-circuit current, maximum-power voltage, and maximum-power current. This reference plot should be measured when the incident radiation is 1000 W/m^2; the cell temperature at that moment must be measured and noted. If the series resistance between the solar cell and the curve-tracing equipment is significant, it also needs to be measured and recorded. The rate at which a solar cell's open-circuit voltage and short-circuit current change with temperature must be ascertained. The above pieces of information are considered constants in our program. The two variables that are input while running the program are insolation and cell temperature. An input insolation of 1 will represent 1 kW/m^2 if the reference curve was in fact measured at that insolation. If the input cell temperature and insolation are equal to those experienced when measuring the curve, the simulated and measured curves should be identical. The following program was written in the BASIC language. The necessary reference curve information must be inserted in the program, and specific graphics commands need to be included to properly scale and label the graph on the CRT screen or mechanical plotter being used.

```
10  ! Voc = reference open circuit voltage (volts)
20  ! Isc = reference short circuit current (amps)
30  ! Vmp = reference max power voltage (volts)
40  ! Imp = reference max power current (amps)
50  ! Tr  = reference temperature (deg C)
60  !  A  = alpha = current temperature coefficient (amps/deg C)
70  !  B  = beta  = voltage temperature coefficient (volts/deg C)
80  ! Rs  = series resistance (ohms)
90    INPUT "INSOLATION (kW/m2) = ", Ins
100   INPUT "CELL TEMPERATURE (deg C) =", Ct
110   C2=(Vmp/Voc-1)/LOG(1-Imp/Isc)
120   C1=(1-Imp/Isc)*EXP(-Vmp/(C2*Voc))
130   V=0
140   Di=A*Ins*(Ct-Tr)+(Ins-1)*Isc
150   Vr=V+B*(Ct-Tr)+Rs*Di
160   I1=Isc*(1-C1*(EXP(Vr/(C2*Voc))-1))
170   I=I1+A*Ins*(Ct-Tr)+(Ins-1)*Isc
180   PLOT V,I
190   IF I<0 THEN GOTO 220
200   V=V+.1
210   GOTO 150
220   END
```

BUILDING A SOLAR CELL ARRAY

The size and power of a solitary solar cell (about 100 cm^2 and 1/2 Wp) are dwarfed by the vast sunlit area of the earth's surface and by the amount of power people need for their homes, work, and play. The earth's surface area facing the sun is almost 250,000,000 km^2, which provides ample space for solar cells. The electric power consumption of a family home during one day ranges between 5000 and 35,000 Wh, which puts a huge demand on a solar cell. To meet both these challenges, photovoltaic cells must be plentiful. Two or more photovoltaic cells can combine their power in various ways when they are harnessed together into an array. However, solar cells interact in certain ways that need to be understood before they can be united in the task of producing electricity; we need to learn their protocol.

A SOLAR-CELL ARRAY EXPERIMENT

Choosing a clear sunny day, we take three identical solar cells outside. To measure their interaction, a voltmeter and an ammeter are brought-along. Pieces of insulated copper wire with alligator clips on either end will be the means of introducing one solar cell to the next. Arranging the cells on a level platform and setting the voltmeter and ammeter nearby, we are ready to perform some tests.

In Chapter 2, we learned that all silicon solar cells have open-circuit voltages of about 6/10 volt. Short-circuit current varies with light intensity, so it is difficult to predict. Let us make a quick measurement by putting the ammeter leads across the cell terminals. At this time of day and year, we read 1 A. For the time being, we will ignore what happens to the voltage and current under a changing load.

Solar Cells Connected in Series

Taking two cells, we connect them in series, or head to tail, with a jumper wire and alligator clips; the positive of the lower cell is connected to the negative of the upper. Taking a voltage measurement across the two cells, we get a 1.2-V reading. Putting the third cell in series with the other two, we measure 1.8-V. This test tells us something most of us already know about batteries: *The voltages of solar cells add when they are wired in series.* The short-circuit current of the three series cells is measured by putting the ammeter across all three. Our ammeter reads 1 A, which is no different than the short-circuit current of one cell. *The current of infinite-shunt-resistance solar cells cannot be greater than the lowest short-circuit current of any series cell.* Much like a series string of Christmas lights, the current in a chain of solar cells cannot be stronger than its weakest link.

Solar Cells Connected in Parallel

Putting our cells in series is only one of two options that lie before us. Why not put them in parallel? Connecting the positive terminals of two cells together and similarly connecting their negative terminals, we have two cells wired in parallel.

A voltage measurement across the two parallel cells gives us a reading of no more than 0.6 V. *The voltage of solar cells in parallel is governed by the individual cells and cannot be any greater than the average of all the single-cell open-circuit voltages.* The cell with the lowest voltage determines this value; since the voltage of different cells in the same environment is very similar, we can say that for all intents and purposes the voltage of cells in parallel is equal to the voltage of one cell. Putting three or more cells in parallel would have little effect on this verdict. If the voltage remains the same for a parallel arrangement, we might suspect that the current is greater. Indeed, a short-circuit measurement gives us a reading of 2 A. Three identical cells in parallel would give us 3 A, four in parallel would put out 4 A, and so on. Therefore, *the currents of solar cells add when they are wired in parallel.*

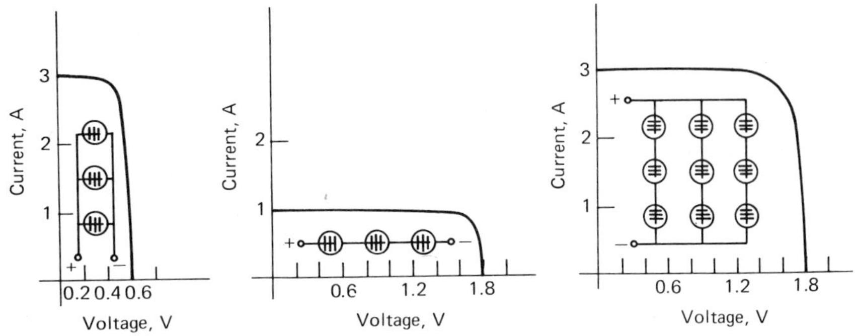

Figure 5-1 **An *I-V* curve of solar cells connected in series and in parallel.**

Array Output under a Changing Load

Photovoltaics normally do not operate at open or closed circuit, for in this state they do not produce any power, and without power we are not pleased. It is time to bring a load into this scenario and put the solar cell to work. By varying the load and plotting our familiar current-versus-voltage curve, we can examine how series and parallel configurations affect its shape.

Beginning with two cells in series, we vary the load across them from open to closed circuit and take data-point readings. Plotting these points on a graph, we obtain the familiar "knee-shaped" *I/V* curve, but in this case it has an elongation along the voltage axis to account for the higher open-circuit voltage. Two cells in parallel produce an *I/V* curve that is stretched high on the current axis because of the increased short-circuit current flow. Notice in Figure 5-1 how the *I/V* curve grows in the horizontal direction as we add cells in series and expands vertically when we put cells in parallel.

Now that we are operating with a load and are dissipating energy, we can hunt for the maximum-power point and compare the power output of these two arrangements. Using the procedure we outlined in Chapter 2, we determine that the maximum-power point of the identical cells has a voltage of 0.5 V and a current of 0.8 A, producing a maximum of 0.4 W. Hopefully, when we put two cells in series, we will double the power, but can this doubling also occur when two cells are put in parallel? Before rushing into an elaborate experiment, let us try to think this one out. If our two identical cells with 0.4 W each are in collaboration, we should expect 0.8 W from both of them. Playing with the numbers, we see that there are a variety of voltage and current measurements that when multiplied together give us 0.8 W. For example:

$$1.0\ \text{V} \times 0.8\ \text{A} = 0.8\ \text{W}$$
$$2.0\ \text{V} \times 0.4\ \text{A} = 0.8\ \text{W}$$
$$0.5\ \text{V} \times 1.6\ \text{A} = 0.8\ \text{W}$$

Our reasoning would lead us to conclude that putting duplicate cells in parallel or in series does not affect the total power, it simply rearranges the voltage and current values. A follow-up experiment to test the power output for a series and a parallel network will confirm our hypothesis.

The solar cell is a small autonomous building block of electric power (when exposed to light) that can be stacked in either the series or the parallel direction to produce a desired voltage and current output. View solar cells, if you will, as the bricks of a house. If we have a fixed number of bricks and we intend to build a one-room box house, there is a specific amount of internal space that we can encompass with this building. The internal space is analogous to the total power output of our solar-cell array. Even though the number of bricks is limited, we can make our room of predetermined size into a variety of shapes; the bricks can be stacked high or spread out level. So it is with our building blocks of solar electric power. The decision whether we should pile our cells up or level them off is dictated by the needs of our load. Once we have determined the load's power requirements (or how big a box house we must build), the next question is, at what voltage and current does the load want its power delivered (does it want a flat or a tall house)? Some loads require a high voltage, while others can settle for less. The resistance and length of wire connecting the power source to a load cause power to be lost in the process of being delivered. The voltage and current output for a solar-cell array are usually arranged in a way that limits the waste of power in the transmission lines (i.e., low current, high voltage).

Array Problems and Their Resolution

Up until now, we always have been working with identical cells. Practical experience informs most of us that no two of anything are exactly identical. Everything down to the smallest snowflake has its own peculiarities. Two cells that are manufactured from the same batch and are the same size will still have slightly different electrical characteristics. The current output is far more erratic than is the voltage. The short-circuit current of cells from the same lot often varies by at least a few percent. If we mix in cells from a different batch, the disparity in cell currents is more pronounced. The causes of this current deviation have something to do with the amount of light reaching the cell surface and the quantity of active cell area exposed (the area that is covered by metal conductor paths is inactive, since light cannot reach the PN junction in that part of the cell). In the manufacturing process, there are small but unavoidable differences in the amount of conductor path metallization or in the density of the antireflective and protective coating. The size of solar cells can vary when some cells have been chipped or manufactured differently. Dust or obstructions can predominate more on one cell than on another. External factors such as the uneven shading of cells in a network can cause some cells to receive less light. All these considerations have a greater impact on photovoltaic current than on voltage. Temperature is the main cause of voltage fluctuation and in most operations, the temperature of different cells in a network does not vary significantly. Differences in the shape of the *I/V* curve from one cell to the next have the nice habit of averaging out to produce a negligible power loss.

When we put hypothetically identical cells in series or parallel, the total power output was equal to the sum of its parts. If each cell had a power output of 0.4 W, connecting two together would give us 0.8 W. The fact that cells are not identical introduces some error. Since current output is the major deviant, the solar-cell wiring arrangement that is most sensitive to cell current will be the hardest hit. Will the series network of cells be more adversely affected by current differences, or could it cripple the parallel organization more? Quite naturally, the "string of Christmas lights," or the series setup, is more vulnerable, since its current cannot be much stronger than that of its weakest cell. The total power in a series-wired string of photovoltaic cells with their usual fluctuations in cell current is not exactly equal to the sum of the power of all the cells; the whole does not precisely equal the sum of its parts. The amount of power lost depends on how much the current output for each cell in the string varies from the average. What we have here is a *mismatch* problem.

Where does the lost power disappear to? The stronger cells with the higher current gang up on the weaker cells in the string and dump their excess power into them. Here we have a common case of bullying; the members of a team that are not able to keep up are harrassed and pushed around. The power absorbed by the sluggish cell has no place to go but into producing heat. The feeble cell usually is able to limp along. However, if the number and size of the bully cells is overwhelming, the weak cell may collapse under the pressure and crack or explode. Here we have a problem called the *hot spot.*

We can get a better grasp of mismatches and hot spots if we do an experiment; perhaps we can figure out a way to minimize the problems. This exploration could prove treacherous, so follow closely. Our test is going to call for cells that have an exaggerated mismatch so that we get pronounced results. Measuring the short-circuit current for various cells, we find two whose currents vary by a factor of 2; the short-circuit current

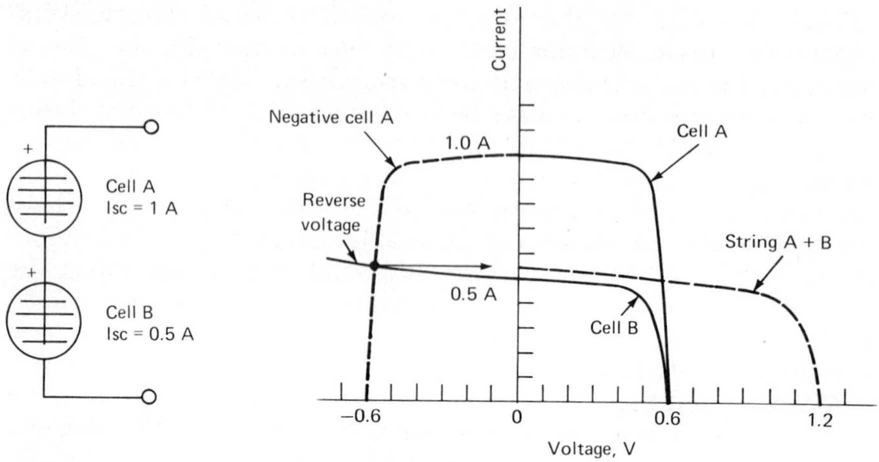

Figure 5-2 The individual and combined *I-V* curves of two mismatched solar cells in series.

of one cell is 1 A, while that of the second is only $^1/_2$ A. Putting these cells in series as shown in Figure 5-2, we find that the short-circuit current for the combined cells is slightly above $^1/_2$ A.

Shorting out two series cells is no different than putting two cells with opposing polarities across each other. The positive of each cell is connected directly to the negative of the other. Portraying the situation on a graph, we find the two *I/V* curves opposing each other on either side of the current axis. At short-circuit current, the higher curve of cell *A* flips over into the negative-voltage quadrant. As shown in Figure 5-2, cell *A* is pushing power into cell *B*; near short-circuit current, the feebler cell *B* stops acting like a power source and gives the appearance of being a load. What kind of load does the weakling cell look like? A passive PN junction is no more than a diode (a device that allows current to flow in one direction and blocks it in the other). Both parts of the PN junction's dual personality, passive diode and active solar cell, come into play in cell *B* as the combination *A* + *B* curve unfolds. On curve *B* we find ourselves wandering off the power-generating portion of a solar cell's curve to the reverse-voltage region of a diode's curve, as shown in Figure 5-2.

The positive terminal of the stronger cell *A* tries to force current into the negative end of cell *B*. Since a diode blocks current flow into its negative electrode, we see that the negative extension of cell *B*'s curve is fairly level and shows little increase in current. A diode's ability to block current, however, is not perfect, and we notice a gradual increase in current as the reverse voltage climbs. If we were to increase the inverted voltage to 30 V or more, we would witness the breakdown of cell *B* under the electromotive pressure. At reverse breakdown, there is a sudden rush of current flow, much like a dam collapsing under the weight of an overflowing river. At this point, cell *B*'s curve would take a sharp turn upward.

Notice in Figure 5-3 how a string of cells can gang up on one weak cell that either has a far lower short-circuit current or is actually broken completely across to make it an open circuit. More cells in a string produce a greater possible reverse voltage across a feeble cell, and the chance of reverse breakdown and a hot spot is increased. A hot spot is far more likely when a cell string is being operated at or close to short-circuit current. A load in the circuit protects the weak cell by limiting the current and dissipating some of the power.

If we could protect the weaker cell from the high voltage its team members are prone to inflict on it, we would alleviate the hot-spot problem. One solution would be to avoid wiring cells in series and to add cells in parallel when more power is needed. This solution is not usually viable, since loads often require a high voltage. A better approach would be to put a device across each cell that could provide an alternative path

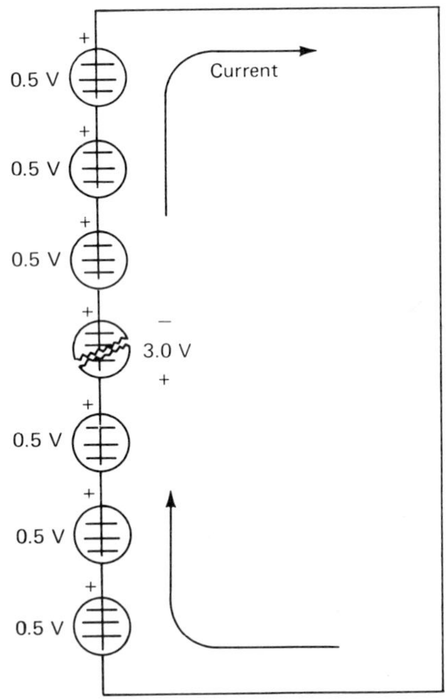

Figure 5-3 **A cracked cell in a series string of cells with a reverse voltage drop.**

for current when a reverse voltage appears. A diode would solve our dilemma. Notice in Figure 5-4 what happens to our weak cell when a diode is put across to protect it. When the reverse voltage rises above 0.6 V, almost all the current flows through the diode, and the weak cell is spared. Since the reverse breakdown voltage of a cell is usually above 10 V, we do not have to put a diode across each cell; it will suffice to put one across a string of 15 or so photovoltaic cells. When a diode is used to protect solar cells, it is aptly called a *shunt* or *bypass* diode.

Now that we have developed a solution to the hot-spot problem, let us focus our attention on cell mismatch losses. Going back to our two series cells with different current outputs, we can run an *I/V* curve for both of them and examine its shape. The short-circuit current for two cells in series is found near the intersection point of the negative *A* curve and the reverse-voltage curve of cell *B*, as shown in Figure 5-2. Follow this point horizontally over to the current axis to find the short-circuit current value. The reverse-voltage curve is not completely horizontal, but rather climbs gradually because a cell's shunt resistance is not infinite. Therefore the short-circuit current of the string is somewhat above the

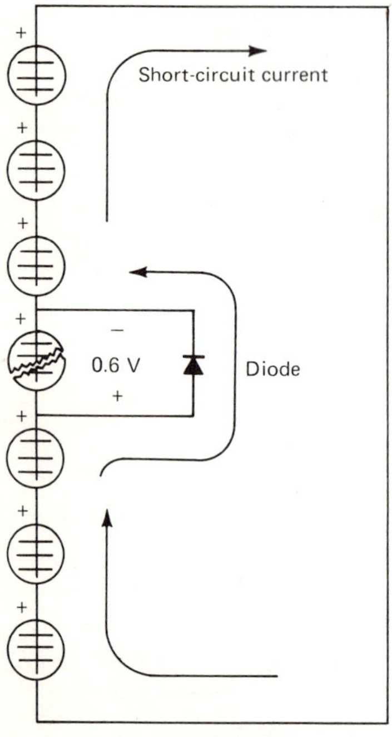

Figure 5-4 **Bypass diode protection around each cell diverts excess current around weak or cracked cell**

short-circuit current for the delinquent cell *B*. Here already we can spy a possible solution to the mismatch problem. If we put more high-current bully cells in series with the weakling, we will operate higher on its reverse-voltage curve. If there are enough strong members in the team, they can lift the sluggish fellow close to their level.

As the load across the two series cells varies toward open circuit, the curve is carried horizontally by the stronger cell *A*. Cell *A* mercilessly forces power into cell *B* until the voltage across the load reaches the open-circuit voltage of cell *B*. At this critical voltage, the feeble cell overcomes its bully; popping out of its passive state as a load, it begins to produce power as a solar cell should. Between 0.6 V and 1.2 V, the $A + B$ curve is being carried by cell *B* and drops off in the usual knee-shaped fashion.

If we could prevent cell *A* from dumping its extra power into cell *B* and direct that current outward, we would reduce any mismatch. The diode we inserted to avert the hot-spot crisis will be of help. At the single-cell level, the voltage will not get high enough to turn the protective diode on. Nevertheless, when we get into large arrays of cells with higher voltages, the diode will indeed shunt excess current.

Another solution would be to connect solar cells in a series-parallel matrix. If we put parallel interconnect wires between successive cells in all the strings, we could shuffle the current around to equalize it. Excess current in one string might be needed in another. Mismatches could also be reduced if we sort cells so that those in a string have as close as possible to the same short-circuit current.

We have found fairly simple solutions to the two major design problems that can hamper an array of solar cells: the hot spot and the mismatch. Diodes across groups of cells prevent hot spots and alleviate some of the mismatch. Cell mismatch can be further reduced by sorting cells into groups with nearly equal short-circuit current and by putting parallel interconnects between cells in parallel strings.

ARRAY CONFIGURATION

Having mastered the art of solar-cell coordination, we now can build up an army of solar cells to deliver power in abundance. Since we will be commanding hundreds if not thousands of cells, they must be organized into manageable groups. A series chain of solar cells is called a *string,* while parallel bunches are named *blocks.*

The first major organization of cells is called the *module,* which is as much a physical arrangement as it is an electrical one. Solar cells, typically ranging in number from 10 to 50, are packaged together in sealed containers that protect them from the elements and make them easier to transport and assemble. The front surface of a module must be transparent and thus is made of glass or clear plastic. The back plate is usually composed of plastic or metal. Solar cells are wired together into various strings and/or blocks and are sandwiched between the transparent front plate and the supporting back plate. Various rubber or plastic sealants are injected to keep the cells snug between their two blankets and to ward off natural erosion from wind and rain. Examples of various techniques for packaging solar cells into modules are shown in Figure 5-5. A module's electrical performance differs from that of an individual solar cell only in that it has a higher power output. Otherwise, all the rules that pertain to a solar cell apply to a module as well. Specifications and cell configurations for various modules available on the market are found in Appendix B.

Moving up through the ranks of the photovoltaic establishment, we

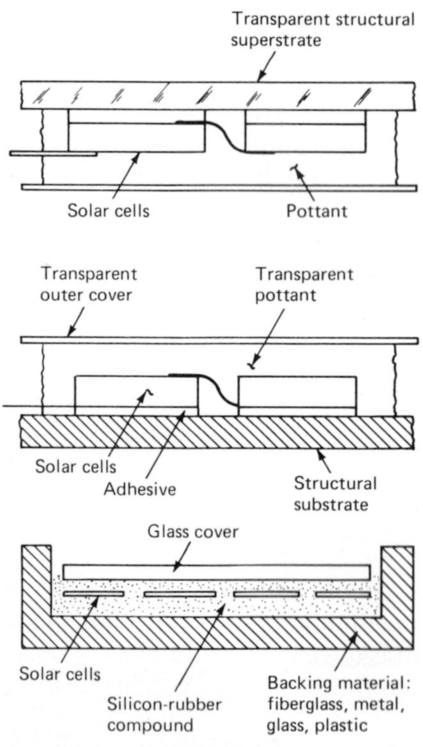

Figure 5-5 **Typical methods of encapsulating solar cells into a module. (From: Monegon, Ltd., *Solar Electricity: Making the Sun Work for You*, 1981, p. 79.)**

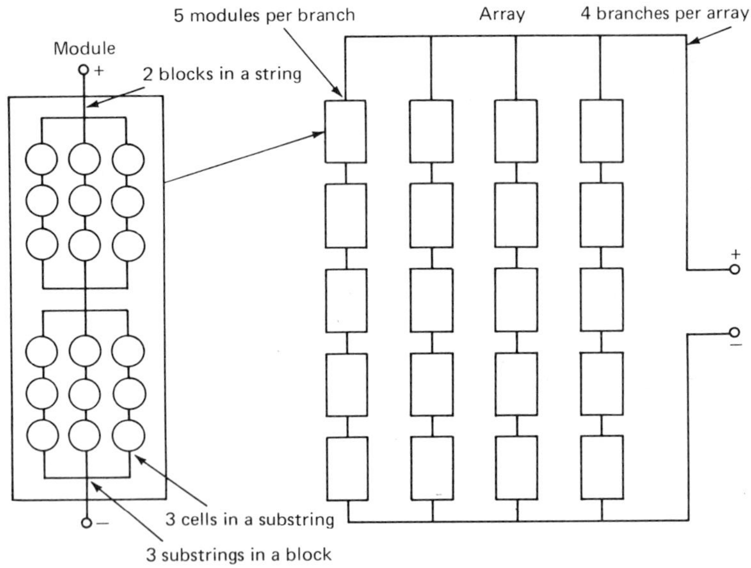

Figure 5-6 Solar-cell configurations in series strings or parallel blocks that build into branches and a total array.

Moving up through the ranks of the photovoltaic establishment, we shall call a series string of modules a *branch circuit.** Finally, the total assembly of solar-cell modules is called an *array,* which is no more than a block of parallel branch circuits. Readers who are bewildered by this scheme will find Figure 5-6 helpful.

This chapter emphasizes wiring solar cells or modules first in series strings and then in parallel branches; the reverse approach is also possible. Solar cells in a module or modules in an array can just as well be wired first in parallel and then in series. In large arrays a branch circuit often is composed of a string of two- or three-module blocks. Troubleshooting, however, generally is easier in a string of modules than in a block, because it is simpler to make a voltage measurement than a current measurement, as will be demonstrated in the next chapter.

To configure photovoltaic modules into an array, we must first know their specifications. A table of the basic PV-module criteria is found in Appendix B. There are essentially 10 pieces of electrically related information that need to be known:

1. Peak rated power at NOCT (PP)
2. Open-circuit voltage at NOCT and one sun *(Voc)*

*The National Electrical Code calls a PV branch circuit a source circuit.

3. Short-circuit current at NOCT and one sun *(Isc)*
4. Maximum-power voltage at NOCT and one sun *(Vmp)*
5. Maximum-power current at NOCT and one sun *(Imp)*
6. *I/V* curve fill factor [FF = (*Vmp* × *Imp*)/(*Voc* × *Isc*)]
7. Module efficiency at NOCT
8. Nominal operating cell temperature (NOCT)
9. Minimum breakdown voltage
10. Cell packing factor (PF = active cell area/total module area)

The first seven criteria can be obtained by taking an *I/V* curve of the module at, or normalized to, 1000 W/m^2 and the NOCT. Fill factor indicates how much series and shunt resistance is present in a module; it should be between 0.6 and 0.8. The nominal operating cell temperature is measured under conditions outlined in Chapter 4; NOCT should be measured when the PV modules are in a mounting configuration typical of the final architecture. The energy conversion efficiency of a PV module should be rated at NOCT. This efficiency accounts for space losses in the module resulting from a cell packing factor of less than one. Often cell efficiency is altered by the way cells are packaged into a module; the designer needs to know the module efficiency and cannot rely on cell efficiency. A module's breakdown voltage indicates the highest array voltage that can safely be handled; an array that has an output voltage in excess of this critical voltage could experience a breakdown in insulation between the solar-cell circuit and the module frame. The cell packing factor indicates how close the cells are arranged; square cells can be packed closer together than round ones. The open-circuit voltage on the coldest and hottest day and the short-circuit current during periods of most intense sunshine should be ascertained for the module and site under consideration in order to know the voltage and current extremes that must be tolerated in the array circuit.

ARRAY ENERGY CALCULATIONS

A photovoltaic-array design is based on a prediction of how much solar-cell energy can be obtained at a particular location and time of year. In Chapter 3, we learned how to determine the amount of solar energy available on a south-facing tilted surface for any location and date. Let us now calculate how much solar-cell energy will be generated from the existing sunshine.

There are six major factors involved in calculating how much photovoltaic electricity can be generated from one day's exposure to a given amount of solar energy. These factors are: average daily solar energy, PV-module efficiency, efficiency temperature correction factor, array packing factor, soiling factor, and total module area. Multiplying all six factors together gives the average daily photovoltaic electric energy.

Monthly *average daily solar energy* data (kWh/m^2 · day) for 20 different sites at various south-facing tilt angles can be found in Appendix C. Insolation data for other sites can be obtained from various publications listed in the bibliography. Most insolation data are taken on a horizontal surface; the equations provided in Chapter 3 will be needed to make the necessary adjustments for an appropriately tilted surface.

A photovoltaic module has a conversion *efficiency* of about 5% to 20%; some 80% to 95% of the solar energy that strikes a module's surface is lost because of reflection or absorption. Module efficiency fluctuates with temperature, increasing as cells become colder and vice versa (see Chapter 4). In order to be relevant, efficiency must be rated at a cell temperature that is typical for the photovoltaic module and mounting architecture under consideration. In order to adjust the module efficiency for one temperature to another, an *efficiency temperature correction factor* can be looked up using the efficiency-versus-temperature plot found in Figure 4-4.

Predicting the cell temperature at a particular site and time will be difficult. An operating cell temperature can be estimated using either of two methods. The nominal operating cell temperature (NOCT) can be measured for the prospective photovoltaic module and its mounting scheme (as explained in Chapter 4). NOCT is measured at an average clear-day insolation of 80 mW/cm^2, and therefore represents the average cell temperature when photovoltaic output power is significantly high. On cloudy days the insolation will remain considerably less than 80 mW/cm^2, but cell efficiency during these periods is not especially important, since the photovoltaic output will be very low. Alternatively, cell efficiency can be adjusted to take into account the average daily ambient temperature at a particular site (NOCT is measured at a fixed ambient temperature of 20°C). The average daytime temperature for 20 different sites is given in Appendix C. Cell temperature is greater than ambient temperature by a factor that depends on the insolation, wind, and module mounting scheme. If a plot of cell minus ambient temperature versus insolation is taken for the particular application, as shown in Figure 4-5, an increase above ambient temperature can be determined for a representative insolation of 80 mW/cm^2. If such information is not available, a fair approximation is that the cell will on the average operate hotter than the ambient temperature by 15° to 25°C for modules that are exposed to ventilation on the back, and by 25° to 35°C for modules that are directly mounted on a surface without back ventilation.

There is a limit to how close to one another solar cells can be put

There is a limit to how close to one another solar cells can be put without causing them to short-circuit. The *packing factor* is the ratio of active cell area to total area; it can be as low as 70% for round cells and as high as 98% for rectangular cells (the packing factor of solar cells in a module will be accounted for in the module's efficiency, and under these conditions need not be considered). There is also lost space between modules when they are laid side by side. This loss of active area should be included in the packing factor. A packing factor of 70% would be represented as 0.7.

Power losses due to dirt accumulation on a PV module's surface can have an appreciable effect on power output, hence we must account for a *soiling factor*. In the dense urban area of New York City, soil accumulation has been found to decrease the performance of glass-covered modules by about 5% to 8% after a six-month exposure. In a rural and agricultural area of Nebraska, pollen and dust have caused about a 3% to 5% loss under similar test conditions. Power losses due to dirt will vary considerably depending on the amount of natural cleansing from precipitation and the amount and type of soiling agents (for instance, the high carbon content of urban pollution is considered the major cause of higher power losses in urban areas). To optimize performance, a solar-cell array should periodically be washed with detergent. The soiling factor is measured in terms of how much power is in fact generated despite the soiling (e.g., a power loss of 6% will result in a soiling factor of 96% or 0.96).

Finally, the total *area* in square meters that is covered by solar cells or solar-cell modules must be known to make the photovoltaic energy calculation. The basic formula for projecting the energy output of a solar-cell array for a typical day in any month is provided in Equation (5-1).

$$\mathrm{PE} = \mathrm{TE} \times \mathrm{ME} \times \mathrm{TC} \times \mathrm{PF} \times \mathrm{SF} \times A \tag{5-1}$$

where PE = photovoltaic energy for one day, kWh

TE = total solar energy on the surface for a typical day of month, $\mathrm{kWh/m^2}$

ME = module efficiency (typically 8% to 20%), in decimal format, that is, 14% = 0.14

TC = efficiency temperature correction factor needed to adjust ME to a cell temperature of NOCT or about 15° to 35°C higher than the average daily temperature at the site (see figure 4-4)

PF = packing factor; the ratio of active cell or module area to total area, in decimal format, that is, 75% = 0.75 (*Note:* ME will already include cell packing factor)

SF = soiling factor; the ratio of energy generated to energy that would be generated if the module cover glass were completely clean, in decimal format (that is, 96% = 0.96).

A = area, m^2

The PV-module specifications supplied by the manufacturer often do not include such detailed information as cell efficiency, area, and packing factor. Typically a module is rated in peak watts at a standard temperature of 25°C or 28°C. If detailed information cannot be obtained, the more basic and approximate equation 5-2 will need to be used.

$$\text{PE} = \text{TE} \times (\text{MO}/1000 \text{ watts}) \times \text{TC} \times \text{N} \qquad (5\text{-}2)$$

where MO = PV-module rated peak power output at specified cell temperature, Wp.

N = total number of modules in the array

These equations do not take into consideration the energy lost in the wire resistance, diodes, battery, inverter, or any other components that might be part of a solar-cell system. The efficiency and losses of such components will be dealt with in the system design chapter.

There are three other considerations that can limit the energy output of a solar-cell array, but are difficult to predict: degradation, shade, and variance from the photovoltaic array's maximum-power point. The designer may wish to include these factors in Equation (5-1) if they prove to be significant and if an effective way of predicting them can be determined.

The rate at which solar-cell modules degrade is not well known as yet, since we are dealing with a new technology that has not been put through long-term tests. The quality of module design, the amount of soiling agents in the air, the temperature extremes that a module is exposed to, and the effect of the sunlight spectrum on a solar cell and its cover glass are all factors that can cause a module to degrade. Based on present data, we can tentatively estimate that solar-cell energy output will decrease at a rate of about 1% to 2% a year.

Shading by trees, buildings, or hills that may be nearby will diminish the amount of solar energy reaching the solar-cell array, but this usually will occur in the early morning and late afternoon hours, when the sun's intensity is at its weakest. During the summer months in the Northern Hemisphere, the sun sometimes sets on the tilted surface before it sets on the horizon. Calculating the precise amount of shading that will occur is difficult; an estimate can be derived by plotting the sun's path for the winter and summer solstices and the southern horizon on one graph, and then approximating the fraction of solar energy that is blocked by obstructions (as discussed in Chapter 3).

The load or storage device that is powered by solar energy may cause the photovoltaic array to operate off its maximum-power point. Solar-cell efficiency, however, is measured at maximum power. A PV system generally is designed to operate as close to maximum power as possible; if power losses are excessive, a different design will be needed, as discussed in Chapter 7.

Photovoltaic modules are bought, sold, and labeled according to their peak power rating (at 1000 W/m^2). It is therefore necessary that we convert an array area figure into a peak power value. This conversion is accomplished using Equation (5-3); it simply requires that we multiply the array area by one sun (1000 W/m^2) and then by the module efficiency and the cell packing factor.

$$\mathrm{PP} = A \times 1000\ \mathrm{W/m^2} \times \mathrm{ME} \times \mathrm{PF} \tag{5-3}$$

where pPP = photovoltaic peak power, Wp/m^2

A photovoltaic array's peak output power and the cell area needed to achieve a certain output level are important pieces of information in the process of designing a photovoltaic array.

Array Design Exercise

As an example, let us design a photovoltaic array that will cover a 30-m^2 area of a south-facing residential roof. The modules that we will use have a peak output power of 60 W with a maximum-power voltage and current of 25 V and 2.4 A and a module efficiency of 10% (all at NOCT = 40°C). Since we are dealing with "module" efficiency, we do not need to know the cell packing factor; we do, however, estimate that about 10% of the active area is lost when the particular modules are placed side by side, thus producing an overall packing factor of 90%. We will first calculate the array's peak power output using Equation (5-2). Multiplying the four factors together, we calculate that if we cover the 30-m^2 area with the available modules, the array peak output will be 2700 (30 m^2 × 1000 W/m^2 × 0.1 × 0.9 = 2700 Wp). Let us increase the array to 3000 Wp so that an even and whole number of modules can be used (which will increase the effective area to 33 m^2). If we need 3000 Wp and each module produces 60 Wp, we will use 50 modules. Examining the load and power-conditioning requirements (which will be discussed in Chapters 7 and 8), we realize that the array output voltage must be about 250 V. This dictates that a branch circuit will consist of 10 modules (10 × 25 V/module = 250 V). One branch circuit will therefore provide 600 W. In order to obtain the full 3-kWp output capacity, we will need five branch circuits (3000 Wp/600 Wp per branch circuit = 5 branch circuits per array). Our array design therefore consists of five branch circuits in parallel with 10 modules in each branch, to produce a 3000-Wp array at 250 V and 12 A.

Let us determine how much solar energy we can roughly expect to obtain from our 3-kWp array in Charleston, South Carolina, on a typical day in March. We must collect the information needed to solve for photovoltaic energy, Equation (5-1). The estimated solar energy available in the spring on a 45° tilted south-facing surface anywhere in the United States is given in Figure 3-10c. These data will be appropriate, since the roof under consideration is in a similar orientation. According to Figure 3-10*c*, the roof should receive about 5 kWh of solar energy on a typical spring day in South Carolina. The module conversion efficiency is 10% (at NOCT = 40°C). The average daytime temperature in Charleston during March (as found in Appendix C) is 15.2°C. The solar-cell operating temperature is about 20° to 30°C above ambient, and therefore will be about 35° to 45°C, which is sufficiently close to NOCT to allow us to use an efficiency correction factor of 1. The packing factor we have established to be 0.9. We will assume that there are no soiling losses, so the soiling factor will equal 1. The array's total area is 33 m^2. Inserting this information into Equation (5-1), we calculate a photovoltaic-array energy output for one average day in March of 14.8 kWh (5 kWh × 0.1 × 1.0 × 0.9 × 1.0 × 33 m^2 = 14.8 kWh). This figure is an average; on a clear day the array will generate more, and on totally overcast days the energy output will be much less. A final design will call for more precise measurements of solar energy, cell temperature, soiling factor, and efficiency.

6

ARRAY PROTECTION AND FAILURE DETECTION

The basic photovoltaic array design presented in Chapter 5 poses certain practical problems that require attention. Large solar-cell arrays installed in various parts of the United States have been operating since 1978. Their performance record to date has been good, despite the fact that they represent a new technology with significant room for improvement. For instance, of the 11,117 modules or about 170,000 peak photovoltaic watts installed by M.I.T. Lincoln Laboratory over the past three years, only about 2.3% have failed so far. The major problems experienced by these photovoltaic arrays can be traced to two causes, stray electricity and module degradation or failure. Stray electricity caused by such external influences as lightning and atmospheric static can enter the PV-array circuit and cause its voltage and current to rise to dangerous levels. Solar-cell module failures occur for a variety of reasons that are largely attributable to weather. A single module failure can cripple a much larger portion of the array and can present a shock hazard to people. Locating a defect in the array can be difficult or time-consuming without an appropriate strategy or troubleshooting feature.

The design problems reported present two major challenges: How can we protect both people and the array from stray electricity, and what kind of troubleshooting strategy can we develop to facilitate repairing the array? Let us explore these problems and develop designs for their resolution.

PROTECTING THE ARRAY FROM STRAY ELECTRICITY

Stray electricity in the atmosphere can be either static or induced; both are capable of entering a photovoltaic circuit and causing the array's voltage to increase. The unwanted electric current created by these atmospheric conditions can be as little as a few milliamperes or as overwhelming as thousands of amperes.

Static Electricity

A solar-cell array is mounted on a structure that will put it somewhere between the sky and the earth, both of which have electrical potentials. The atmosphere acquires varying quantities of static electricity from the earth, much as our bodies pick up excess charges when we walk across a thick carpet. Certain weather conditions, such as cumulus cloud formations or thunderstorms, increase the level of this static. During an electrical storm, the imbalance in charge between the sky and the earth has become so great that the insulating barrier of air between them momentarily breaks down and the excess electrons jump from the sky to the earth (or vice versa) in a flash of lightning. Lightning strikes the object that stands the highest off the ground because it provides the path of least resistance to ground.

Protection against lightning is a black art that has many different theories and practitioners. The amount of time and money spent on warding off the jagged bolt from the heavens will depend on the frequency of thunderstorms and the vulnerability of the site. The effectiveness of lightning rods in preventing and diverting direct strikes is debatable; nonetheless, it probably would be a good practice to put two or three lightning rods above the array and to connect them directly to a metal rod that sticks at least 8 ft into the ground. Water plumbing pipes sometimes serve as ground rods, but usually it is best to drive a hardened copper stake with a diameter of about 3/4 inch into the ground. The connection between the lighting rod and the ground rod is made with a thick metal cable (American wire gauge no. 000) that can handle the huge current that might flow through it. A lightning rod places the ground potential above the array and will divert any current from a lightning bolt away from the solar-cell circuit.

A second precaution is to connect the array frame and support structure (if it is metal) to earth ground. This provides a quick detour to ground if part of a lightning bolt should jump from the lightning rod to the frame. Without these protective measures, a lightning bolt could enter the electric wiring of the solar-cell array on its way to ground and wreak all sorts of havoc.

Human safety, a crucial consideration, is well served by grounding the array frame and support. A person who accidentally experiences an electric shock from a solar array may just be a little startled or may be dead. A properly functioning array should have a voltage only at the module and array terminals, and these points normally are well concealed. However, during a malfunction, the insulation between a live wire and any adjoining metal or between a solar cell and the metal frame of its module can break down. At this point, the metal frame or support structure has a voltage on it that could be dangerous. A person touching this metal could experience an electric shock. To keep our lives as uneventful as possible, we connect all exposed metal that could become electrically "hot" to ground. We thereby short out any stray voltages between the array frame and the ground upon which we usually stand. A similar precaution is taken with many household appliances. The metal case of an electric stove, or a refrigerator, or a hand drill is grounded to short out any "live" wires that might accidentally come in contact with the case.

Static electricity in the atmosphere can have the milder impact of causing the array's output voltage to float high above earth ground. Up until now, the photovoltaic array has not had a solid and safe point of electrical reference; the array voltage has been allowed to float relative to the world around it. In previous chapters, we have always measured a solar cell's or an array's voltage with respect to its negative terminal. If we measure the voltage from the earth to either of the two terminals of a floating array, the readings can be any of a wide range of voltages that reflect the weather conditions and the amount of static electricity in the air. This voltage generally would be near zero, but under certain atmospheric conditions it could rise into the thousands of volts.

Our feet and our house are situated on the ground, and it is dangerous to have the array voltage occasionally floating above our point of reference; the insulation btween the circuit and ground may break down, or people and equipment may be damaged when the array floats to a high voltage. We must establish a solid electrical foundation. As the largest conducting body and as the surface upon which most of our activities take place, the earth is used as an arbitrary zero reference of potential and as a common return for most electric circuits. One of the wires coming from the electric utility to any building is always grounded. Earth ground therefore becomes a reference from which all voltages in a circuit are measured. To anchor the solar-cell array to ground, we should make an electrical connection between part of the array and the earth ground rod. Typically the negative terminal of a PV array is grounded; however, a designer may choose to ground the middle of the array circuit (i.e., center tapping) to minimize the voltage drops to ground and thus reduce the stress on the electrical insulation and components.

Grounding the array's negative terminal eliminates the problem of the array voltage floating high but introduces a new problem. If a malfunction in the array causes a cell or wire to short to grounded metal, current will flow through ground back to the negative terminal of the array. Although this short-circuit path facilitates troubleshooting (as will become evident in the next section), it creates a fire hazard that must be avoided. When the defective short to ground somewhere in the array circuitry is not complete, electricity may jump across the divide in the form of an arc that dissipates heat and can easily start a fire if there is any flammable material in the vicinity. There are many places in an array where a short to grounded metal is possible, and therefore there is a real likelihood that a short to ground will occur sometime during the lifetime of a photovoltaic array. The potential for arcing and fire will be prevented if there is no solid return path to the array circuitry from ground, and hence if the array's terminal is not grounded.

We are faced with a dilemma; we want to eliminate both the chance of arcing and fire and the chance of a static-induced high array voltage, but apparently we can only prevent one or the other. A compromise solution that will avert both hazards involves inserting a connection between the array's negative terminal and ground that is somewhere between a complete short and an open. A resistor placed between these two points can be sized large enough to prevent current from flowing and arcing from occurring when a short to ground appears in the array, and yet simultaneously be small enough to bleed off static charge when the array's static-induced voltage floats high. A resistor that allows about 10 mA of current to flow in the event of a short (R = array open-circuit voltage/10 mA) will be adequate for this purpose. A short to ground can be detected by sensing the voltage across this resistor.

In the later troubleshooting section of this chapter, we will assume that the negative terminal of the array is grounded because it is easier to detect problems that way. In an array without a solid ground, this connection can be made temporarily when the array is tested, then removed during normal operation.

Induced Electricity

The second source of stray electricity is the current induced in a circuit when moving lines of a magnetic field cut across a conductor or wire. Magnetic fields of varying strength can be found in the atmosphere and are capable of traveling through most materials, including air.

The source of magnetism and induced current is explained in most physics textbooks. According to Oersted's fundamental law, electric current flow generates a proportional magnetic field that encircles the current path. If the amount of current fluctuates, so does the amount and strength of the magnetism. Common examples of current paths that generate a magnetic field are lightning bolts and alternating current in utility lines and radio transmitting antennas. Faraday's law approaches this phenomenon in reverse: Suppose a magnetic field were to pass across a wire; would the converse be true? Indeed so; current is induced in a wire that is under the influence of a changing magnetic field. The size of this induced current is determined by the rate at which the current flow changes. A 100,000-A current often flows in a fraction of a second when a lightning bolt strikes the earth. The magnetic field produced at these moments is prodigious.

The current induced by lightning bolts can be dangerous. A large current surge through a wire's resistance translates into a voltage spike that can destroy the wire, switches, meters, and any other equipment attached to the power lines. To protect the circuitry against these intrusions, we have to install electrical "release valves" that safely divert excess current to ground. The two-terminal semiconductor device called a *varistor* can serve this purpose well. Varistors are installed from the positive and negative leads to ground in both utility and photovoltaic circuits. The resistance between the two terminals of a varistor remains an open circuit until the voltage across them exceeds a designed breakdown level. At breakdown, the resistance drops to near zero, and whatever current is available is shunted away to ground; the resistance returns to its open-circuit state when the voltage drops below the critical level.

The array in Figure 6-1 is protected against both types of stray elec-

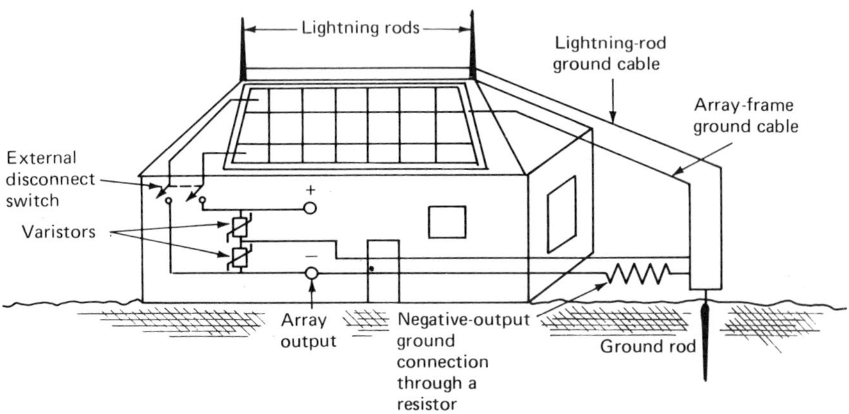

Figure 6-1 A roof-mounted array with protection against static and induced electricity.

tricity, static and induced. Direct-hit lightning bolts will be deflected from the array by the lightning rods mounted above the solar-cell modules and the cables connecting these rods and the metal array from earth to ground. The grounded negative end of the array prevents static electricity from causing the array voltage to float high. The pair of varistors from either terminal of the array to ground shunts away dangerous amounts of electricity induced in the power lines by indirect strikes of lightning. As a final safety precaution, an external disconnect switch is put between the array and the internal house wiring.

ARRAY TROUBLESHOOTING AND PROTECTION STRATEGIES

The Array Problem

The major task of ensuring the continuous operation of a solar cell is insulating it from nature's elements. A solar cell that is protected from vibration, stress, and the chemical and physical effects of air, water, and dirt could function almost indefinitely. Ultraviolet radiation from the sun has a deleterious effect on the photovoltaic's crystalline structure; however, most ultraviolet rays are filtered by the earth's atmosphere. Consequently, a clean and well-protected terrestrial solar cell degrades at a very gradual rate, and will maintain close to the rated output for at least 20 years, if not longer.

To keep out the weather's eroding and corroding influences, solar cells must be sealed between a transparent front plate and a supporting back plate. Choosing the design and materials for a solar-cell module is no small task. Common fabrication materials such as metal, glass, and plastic have conflicting expansion coefficients (changes in temperature cause the substances to expand and contract at different rates). A problem arises when one material in a module expands faster than another; stress is put on the solar cell and its interconnects that may cause them to crack. A crack across the main conductors of a solar cell will open-circuit the string. Sometimes these cracks intermittently open or close the circuit as the crack expands and contracts with changing temperature. The plastic sealant holding a module together can lose its adherence after extended exposure to the weather, and this can cause the top and bottom plates to delaminate. If moisture gets into the module, the solar cell and its interconnects will corrode, and the moisture will provide a path for current to leak away. A module's transparent front plate is threatened by large hailstones and by rocks thrown by mischievous youths, which can also cause cells to crack or current leakage paths to develop.

Most of the serious problems that a photovoltaic array can experience will fall into two categories: an *open circuit* or a *short circuit.* A faulty open circuit involves an interruption in the normal current path such that the circuit is broken, while a short-circuit fault provides a low-resistance detour in the circuit that diverts current away from its productive course.

The Short-Circuit Fault

A short-circuit usually occurs between part of the circuit and ground. Most commonly, the insulation between the grounded metal frame and support structure and the solar-cell electrodes or interconnecting wire breaks down. When the negative electrode of the array is connected to ground, a current ground loop is created in which current is diverted from the short-circuited point in the circuit through the ground and back to the negative terminal of the array. In Figure 6-2 we see how current is lost in this manner. A short circuit can also occur between two or more places within a circuit. A common example would be failure of the protective shunt diode across a module in such a manner that it short-circuits the module and effectively removes it from the array. Figure 6-3 shows how the string current bypasses the short-circuited module. When part of the module string has been short-circuited, the output voltage will be lower than normal.

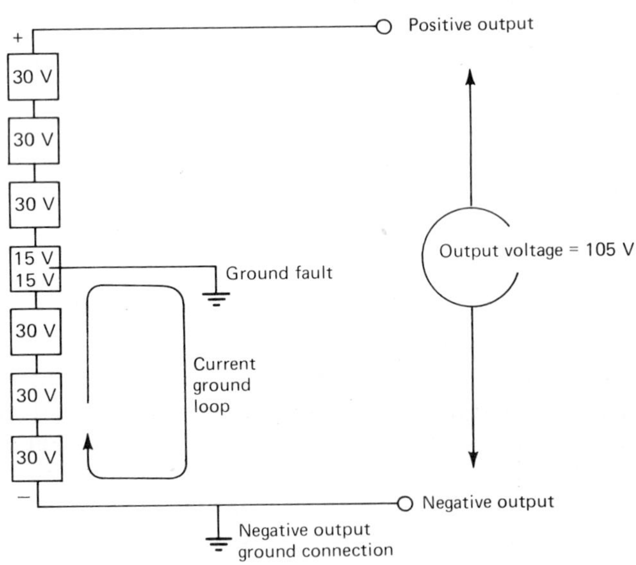

Figure 6-2 Current ground loop created by a ground fault.

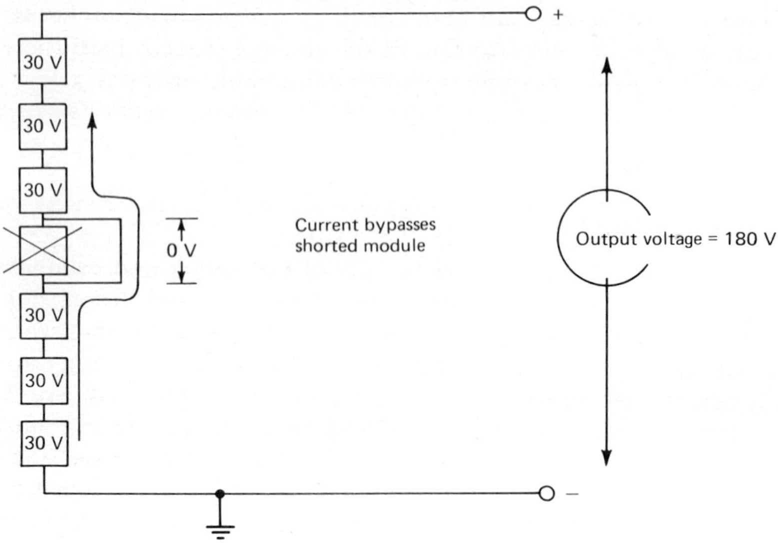

Figure 6-3 **A shorted module contributes no voltage or power to the array output.**

The Open-Circuit Fault

An open-circuit fault generally is caused by a broken wire or terminal or by a broken cell that has cracked completely across its main conductor paths. When the open circuit occurs inside a module and other modules are wired in series, the protective shunt diode diverts current around the broken unit, as demonstrated in Figure 6-4. The output voltage for a module string with an open-circuited module will be lower than normal because there is no output voltage from the defective part. If the open circuit occurs outside the module and somewhere in the wiring, the current path is broken, and no current or voltage will appear at the output.

A conditional open circuit occurs when a module has a lower than normal current output or has a cell with a crack across its main conductor path that intermittently makes or breaks contact. Any factor that obstructs the sunlight reaching a module will reduce its current output. Uneven amounts of dirt or shade or a chipped cell that has a small surface area will produce this effect. A low-current module in series with other normal modules will produce power when the string current is below its limit. However, when the string current exceeds the low-current module's maximum output capability, the module will be reverse-biased and will act like an open circuit. At this moment, the protective shunt diode will divert excess current around the poorly performing module.

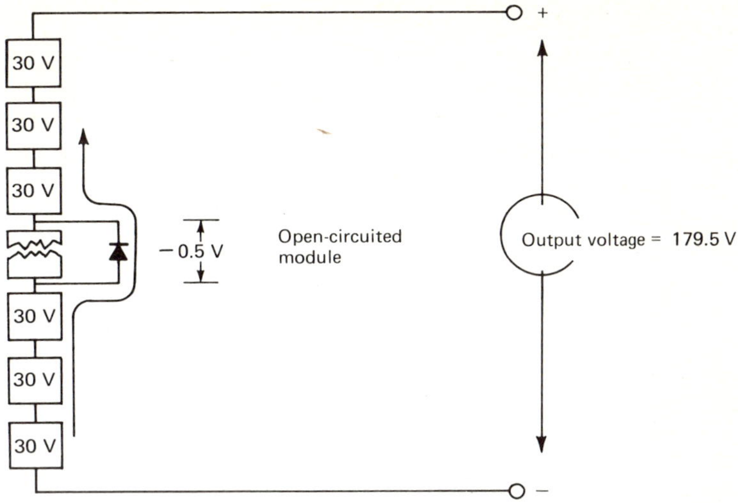

Figure 6-4 **A bypass diode shunts current around an open-circuited or low-current module.**

A low-current module will not be evident when the string output is measured at open- or short-circuit current, but will show its effect only when the string has a load on it. Another type of conditional open circuit occurs when a solar cell is completely cracked across its main conductor paths and yet intermittently makes or breaks contact depending on the temperature-induced expansion and contraction of the cell material. The protective diode turns on and shunts current whenever the cracked cell open-circuits the module.

The Development of Short- and Open-Circuit Faults

Erroneous values of resistance in the array circuit between a short and an open in many cases are transitory and gradually develop into a short or an open. Theoretically, a short is a zero-resistance path, while an open circuit has infinite resistance. In practice, however, zero and infinity never are achieved; in our case a short will be considered to have a resistance of a few ohms or less, while an open circuit will have one million ohms or more. When a short fault is only partial, so that the resistance is greater than a few ohms, it generally is called a *leakage path.* When a circuit is partially broken open and there is a higher than normal resistance in the conductor, it will be called *series resistance.*

A dead short or a complete open does not dissipate any power, since in the first case the voltage is near zero and in the latter case the current flow is nil (power = current × voltage). However, when a leakage or series resistance occurs in a circuit, the current through the resistance multiplied by the voltage across it indicates the power being spent in the trouble spot. When power is dissipated, heat is produced. If the power loss and heat become acute, melting will lead to fusion or breaking in the circuit.

Insulation between the solar-cell circuit and ground often breaks down gradually, first becoming a high-resistance path to ground. Small amounts of current leak away. As current flows through the weak spot in the insulation, power is dissipated ($P = I^2R$), causing the insulation to heat up and further degrade. A high resistance eventually may lead to a low-resistance path, much as a crack in a dam usually leads to a collapse in the barrier and an ensuing flood. When the insulation has broken down completely, the metal conductor and the grounded metal structure can become fused together and remain a low-resistance path or short to ground.

Approaching the problem from the other end, a conductor that is designed to handle large quantities of current in a solar-cell circuit may crack or degrade, so that its resistance increases. The increased resistance in series with the circuit heats up as more power is dissipated and may eventually burn away until it becomes an open circuit. This series-resistance fault or weak point in the circuit strains under the high current much like a weak joint in a pipe that bursts when the water pressure becomes excessive.

Series resistance or a leakage path is often the precursor of an open or a short. The power spent and heat generated in a leakage path gradually destroy the failing insulation and fuse metal portions of the circuit together or to ground, producing a short. Conversely, the heat generated in a series-resistance fault slowly causes the conduction path to burn up and break into an open circuit.

A Problem-Solving Circuit

A short or an open circuit will cripple the array's performance, and therefore it is important that there be a technique for discovering the fault and repairing it. There are countless places in an array where a problem can crop up; the trick is to design the array so that the type and location of a problem can be easily identified. A little forethought at this point will save a great deal of time and aggravation later on.

There are two diagnostic techniques that can be used to monitor the array's performance: voltage and current readings and *I/V* curve plots. Voltage and current can be measured using a simple and inexpensive multimeter, while *I/V* curves must be plotted with more sophisticated and expensive equipment. Examples of both types of instruments are shown in Figure 6-5. On large arrays, measuring voltage and current in many parts of the photovoltaic circuit can be tedious and time-consuming. A photovoltaic *I/V* curve tracer can simplify large jobs enough to be worth the additional cost. Both diagnostic methods require that the array-wiring scheme be designed in such a way that the multimeter or curve tracer can have easy access to various electrical junctions in the array.

A basic array-wiring scheme is to connect modules in series, then bring the positive and negative wires from each string or branch circuit into a major junction box, where they are connected to blocking diodes and a parallel bus bar. The bus bars carry the total array voltage and current and are the array's output. Tests can be performed on the entire array or on any branch by disconnecting the branch circuit from the bus bar. This wiring configuration is simple and inexpensive, but poses some safety problems. A photovoltaic array cannot be turned off at will; during daylight hours it produces electricity, and any maintenance performed on it requires special precautions. Troubleshooting is made safer and easier if some devices that simplify taking independent voltage, current, and *I/V* measurements are added to the array circuit. Additional circuit protection devices must also be added to the array to prevent certain malfunctions.

Our array troubleshooting design will start with one independent branch and build up to its connection with the bus. First, the branch power lines will have a pair of varistors from the positive and negative legs to ground, as described earlier under stray electricity. Troubleshooting is simplified if quick voltage and current measurements can be made with little danger to life and limb. The subarray voltage will be monitored by installing a voltmeter across the power lines. The current is best measured by putting a shunt of known resistance in series with one of the power lines and measuring the voltage across the shunt.

A current shunt enables us to measure the current without actually putting a meter in the circuit. A shunt's resistance typically is very small (between 1 and 300 mΩ) to keep the power loss in it to a minimum. The current is determined by dividing the voltage across the shunt by its resistance. A meter wired across the shunt measures the voltage and scales it to read current. This remote sensing of current makes the physical wiring of the circuit easier and safer. The wire coming from an array will handle large amounts of current, and therefore it must have a relatively large diameter to keep the resistance low. On the other hand, the wire going to a meter is of small diameter because the full-scale current is very low, in the neighborhood of a few milliamperes or less. We want to keep the amount of large-diameter wire to a minimum

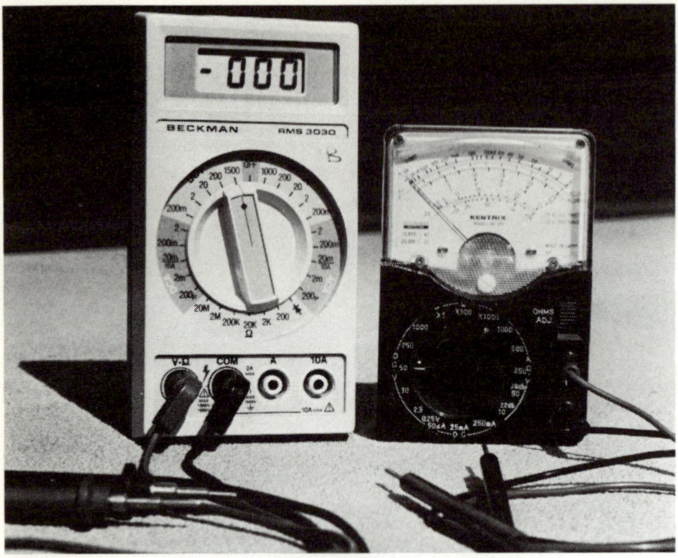

(a)

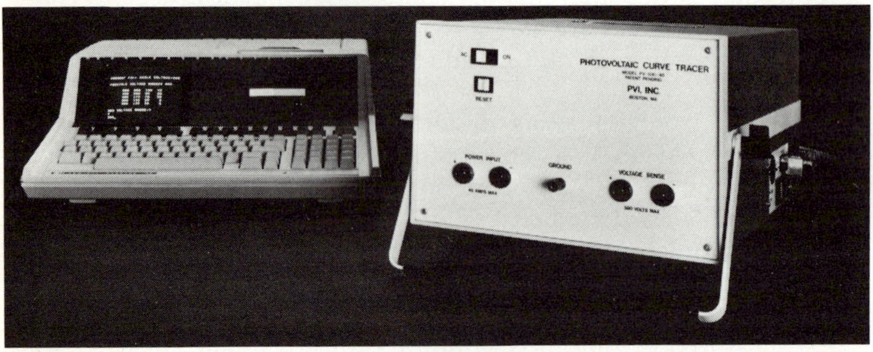

(b)

Figure 6-5a and b **Two PV-array testing instruments: (*a*) an analog and digital multimeter (photo by author); (*b*) a PV-array *I-V* curve tracer (courtesy PVI, Inc, Boston, Mass.).**

because of cost and space considerations. The high-power wires also must be kept enclosed and free from potential hazards. Routing power lines directly through a panel-mounted ammeter can increase the likelihood of arcing to ground. For a similar reason, it would be safer to put the shunt in the negative leg of the branch output. Since the current coming out of the branch should equal the current going in, it matters little at which end we sense the current. The negative end has zero volts

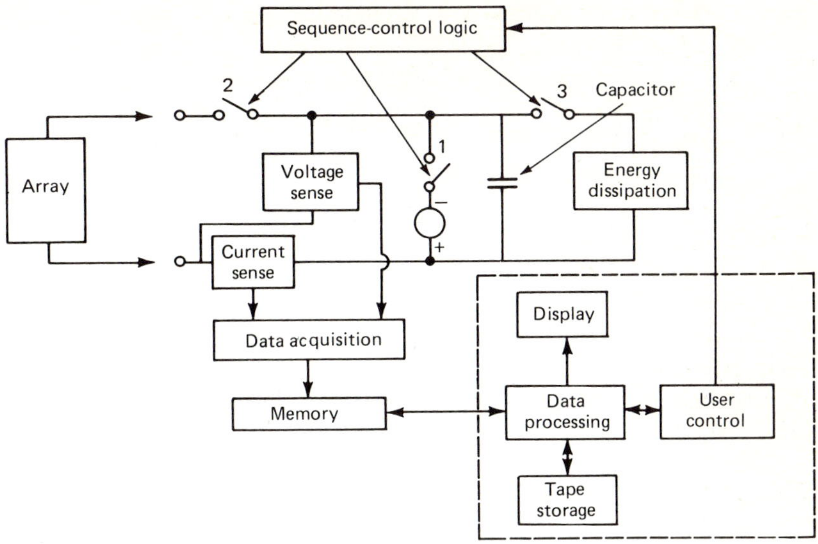

Sequence

1. Precharge capacitor in reverse polarity.
2. Briefly connect PV array to capacitor and collect data during charging period.
3. Discharge capacitor in preparation for new *I/V* curve measurement.

Figure 6-5c **block diagram of *I-V* curve-tracer circuit. (T. H. Warner and C. H. Cox III, *A High Power Curve Tracer Employing a Capacitive Load,* M.I.T. Lincoln Laboratory, Lexington, Mass. pp. 3–4.)**

to ground and would invite fewer hazards if the ammeter were shorted to ground.

As an additional safety precaution, a fuse (with about a 1/16-A rating) will be placed in the wire running from the positive leg of the array to the voltmeter. This fuse will prevent large amounts of current from flowing into the voltmeter should it short to ground accidentally.

Easy and safe access to test points in the branch output will facilitate making other measurements. Following the two meters, wires capable of handling the full current will run from both the positive and negative leads to a well-insulated and protected connector on the front panel. This connection will enable us to perform tests such as making a current-versus-voltage curve or measuring leakage to ground.

Our major troubleshooting strategy involves installing two switches in the power lines coming from the branch circuit, as shown in Figure 6-6. The first switch will enable us to short out the branch by connecting the positive and negative leads together. With this "short" switch closed, the ammeter will indicate the short-circuit current. A second switch will make it possible to float the branch by disconnecting it from earth ground and the main bus. With this "disconnect" switch in the open position, the voltmeter will read the open-circuit voltage. The short and disconnect switches also serve important safety functions in that they can manually remove the branch-circuit power from the main bus in the event of serious electrical problems at the array output. A single-pole switch will be used to short the branch, while a double-pole switch will disconnect it. The number of poles indicates the number of connections that simultaneously can be opened or closed. Since floating a branch entails opening both the positive and the negative leads at the same time, a double-pole switch is needed. Under normal operating conditions, the short switch will be open and the disconnect switch will be closed.

The electric energy from a solar-cell array usually must be stored and/or converted from direct current to alternating current before it is put to use. The various branches are connected to the bus, and the bus is attached to a battery and/or an inverter. There is a real danger that the large reservoir of energy stored in a battery or available in the utility grid (to which many inverters are attached) could accidentally feed back into the array. Especially at night, when the array has no output voltage, the possibility of a reverse current flowing into the array has to be dealt

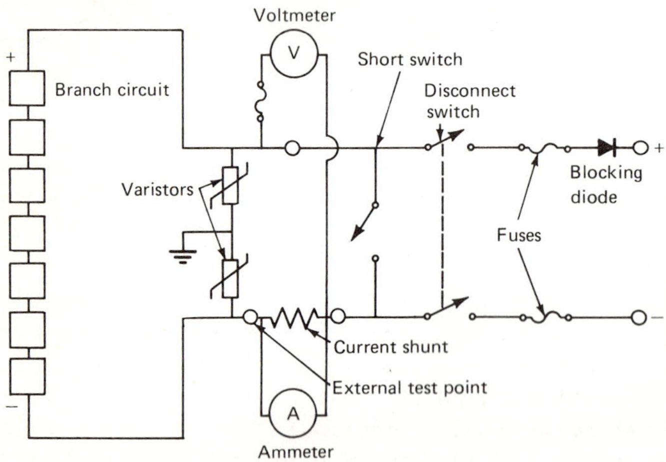

Figure 6-6 A branch circuit with protection and troubleshooting features. (Adapted from the M.I.T. Lincoln Laboratory prototype design at the Northeast Residential Experiment Station.)

with. There is also a danger that a short in one branch of a photovoltaic array will draw current from other branches in parallel. A diode placed in the positive lead and a fuse inserted in both the positive and negative leads coming from each branch will alleviate this potential hazard. The cost of a few diodes and fuses is trivial compared to the damage or power-loss expenses that could be incurred if the aforementioned problem were to occur.

The blocking diode will be positioned in the positive lead in such a way that its positive electrode is facing away from the array, as shown in Figure 6-6. It will allow the current to flow from the solar cells to the house or load, but will prevent any current flow back into the array. This diode also will prevent the short switch from shorting out the battery or inverter input when we use it in our troubleshooting procedure. The peak-inverse-voltage rating of the diode should well exceed the open-circuit voltage of the array, and the forward-current rating must be greater than the branch's maximum current.

A fuse is a piece of metal conductor that is designed to melt at a specific current. It is an intentional weak link in the circuit that will interrupt excessive current flow before the wire or some other vital part of the circuit is destroyed. The fuse and interconnect wire should be rated to handle about twice the peak branch current (to allow for instances of natural solar concentration). The DC voltage and current rating of the fuse must exceed the branch's voltage and current extremes to prevent any arcing across the fuse terminals. Fuses act as backup protection should the diode fail and allow reverse current to flow in the array or the varistors be unable to handle a large current flow during a lightning-induced surge. For these applications the fuse should have as fast a response time as possible and be rated for DC.

All the branch circuits are connected together in parallel after their respective diodes on a positive and negative bus bar, as shown in Figure 6-7. The bus bar usually is made of solid copper, with threaded holes where screw terminals are inserted. Rather than having a separate voltmeter and ammeter for each branch, we can install a pair of meters for all the branches and use a three-pole rotary switch to move the meter probes to the branch of our choice. The array main bus will also have a pair of varistors from the positive and negative bus to ground. A panel-mounted voltmeter will be connected across the two bus bars, and a current shunt with a scaled panel-mounted ammeter will be installed at the output end of the negative bus bar. The negative bus will be connected to earth ground by a cable that is capable of handling considerably more than the maximum array current. (This ground connection may be made at a later point in the power-conditioning circuit in certain

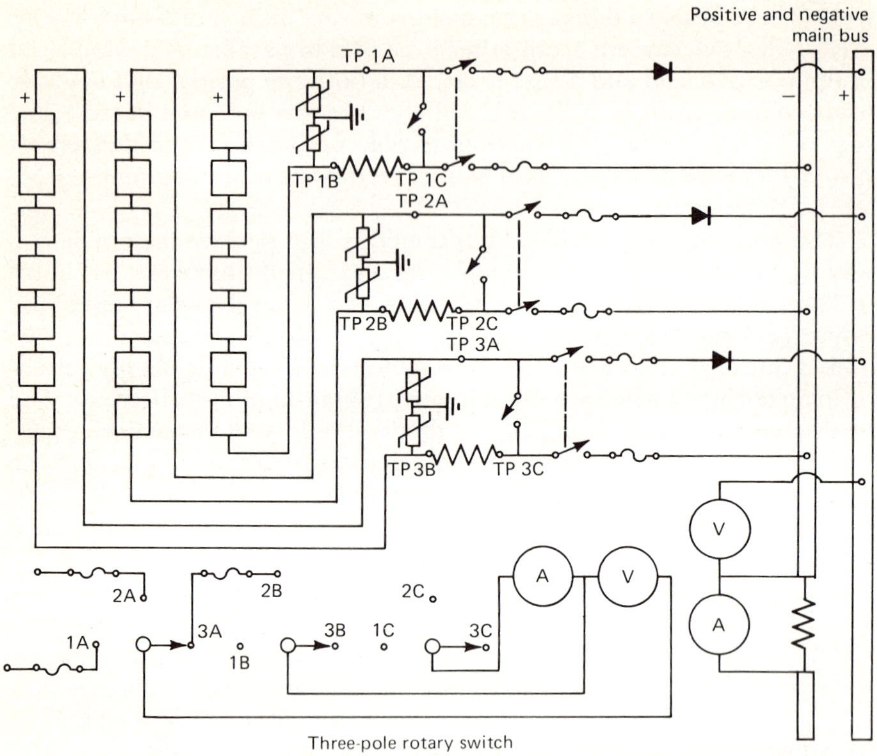

Figure 6-7 **The total array with protection and troubleshooting features.**

designs discussed in Chapter 7.) A complete junction box with the circuitry described in the section is shown in Figure 6-8.

The design we just have developed will make it possible to isolate a problem to the branch level. How do we locate a defective module within the branch? Even though certain malfunctions can be pinpointed by making measurements at the branch output, troubleshooting is greatly simplified if voltage and current measurements of each module can be made. Making these measurements requires access to the module terminals. If the module mounting scheme makes electrical contact with each module impossible, we could consider bringing wires out from the junctions between modules and running them to a test box. These wires should be capable of carrying the full array current and will make it possible to test modules individually. A test box can be a hazard if it is not properly wired and insulated or if it is used by an uninformed person. The cost and danger of such a test box make it an unattractive option.

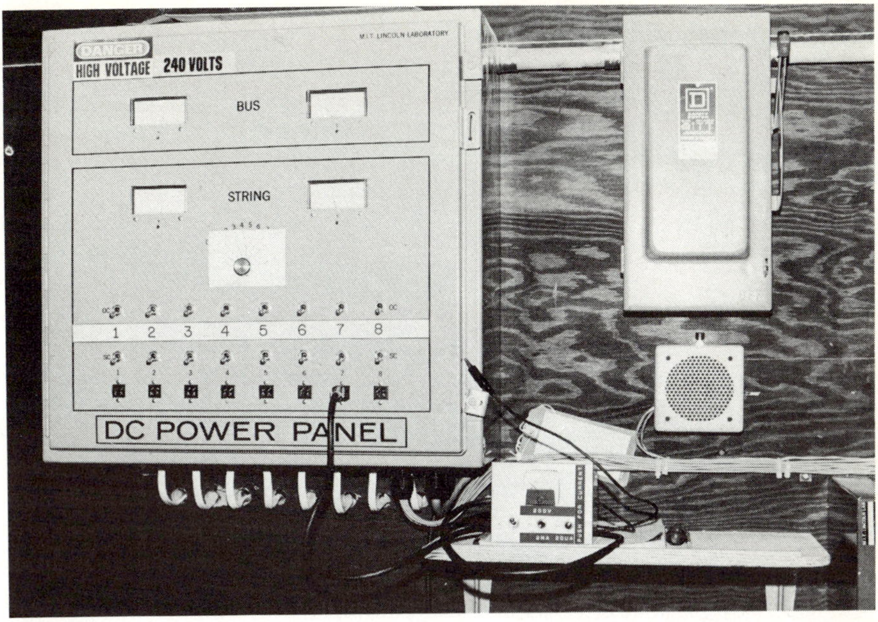

Figure 6-8. **A junction box for a photovoltaic-array output with troubleshooting features: disconnect (OC) and short (SC) switches, voltmeter and ammeter access to each branch, and connectors for external electrical contact with each branch. (M.I.T. Lincoln Laboratory prototype at Northeast Residential Experiment Station in Concord, Massachusetts; photo by author.)**

The selective shading troubleshooting technique discussed later will be a better way of testing modules without exposed terminals.

Solving the Array Problem

The array has been wired in a configuration of switches and meters that simplifies troubleshooting. Let us now put this design to work to discover basic problems. A well-made module and properly installed array should experience few if any problems, so the first step is to check whether a problem indeed exists.

A quick and simple array test is to compare the voltage and current output of each identically wired branch to see if they are similar. With the disconnect switch closed, the branch voltmeter should read half a volt higher than the array voltmeter because there is a voltage drop across the diode between these two points. The branch voltmeter will

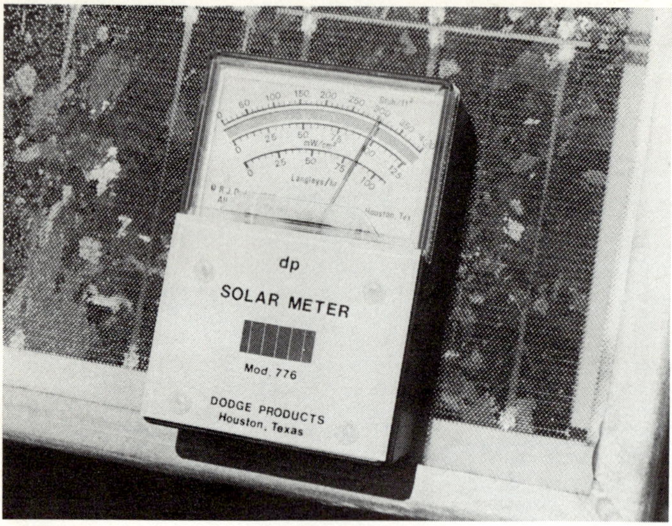

Figure 6-9 **Measuring solar insolation on the array by arranging a light meter with the same orientation. (Photo by author.)**

register an open-circuit voltage when we open the disconnect switch. If we close the short switch, the ammeter will measure the branch's short-circuit current. If the voltage or current of one branch is lower than the norm by more than about 10%, a problem is indicated.

A more accurate and objective test of the short-circuit current of any branch involves first calculating the expected current from the given insolation. We can calculate the expected short-circuit current at peak power if we know the specifications for the modules in a branch. Using what we learned in a previous chapter, we add the current of modules in parallel and take the lowest current of modules in series. The current figure we deduce is rated at peak power; it is unlikely that the insolation at the time of our test will be exactly 100 mW/cm^2. We must adjust the short-circuit calculation we just have made to the amount of insolation that the array is actually experiencing. At this point a simple solar-cell light meter would provide a convenient method of measuring the amount of sunlight. Other insolation-measuring devices, such as a pyranometer, have greater accuracy but also cost more. Place the light meter facing toward the sun at the same position and angle as the array and take a reading, as shown in Figure 6-9. Dividing the meter reading, scaled in milliwatts per square centimeter, by the peak power figure of 100 mW/cm^2 produces a ratio of present to peak power insolation. Multiplying the peak power short-circuit current of the branch by this ratio will give us the expected current at the present insolation.

A discrepancy of greater than 10% between the expected and actual short-circuit current indicates that the branch is suffering from some sort of problem. Visually inspect the branch in question to see if snow, dirt, or shade is covering the modules. When possible, examine the modules to determine if any of them have broken transparent covers or other obvious defects. If the problem continues to be a mystery, check whether there is a short or an open fault in the branch.

Detecting Shorts to Ground Using a Multimeter

A short or leakage path to ground will register on an ammeter that is placed between either lead of a floating branch and ground. Notice in Figure 6-10 how the ammeter completes the current loop and provides a path for the leakage current. A variable-scale ammeter that is capable of handling the full branch current is used for this test. Disconnect the branch from the main bus and keep the short switch open. Set the ammeter on its highest scale, and carefully put the two meter probes from the positive test point to ground. Should there be a short in the branch, the full branch current will be able to flow in the ammeter, and arcing will occur as we make and break the contact with our meter probes; by all means do not let your body become a path for some of this current. Switch the ammeter scale down until a significant current

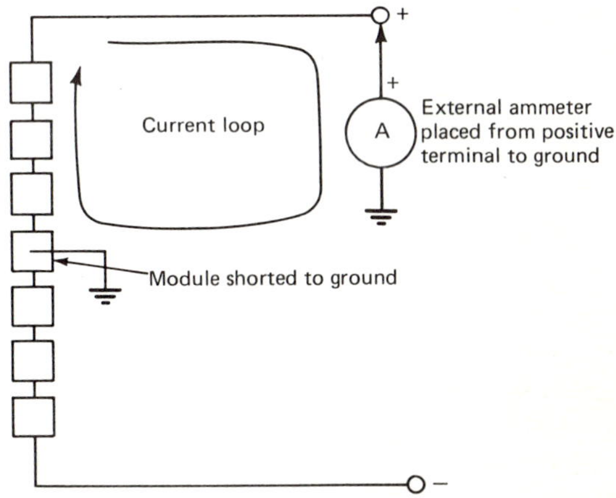

Figure 6-10 **Testing for leakage or a short to ground with an ammeter.**

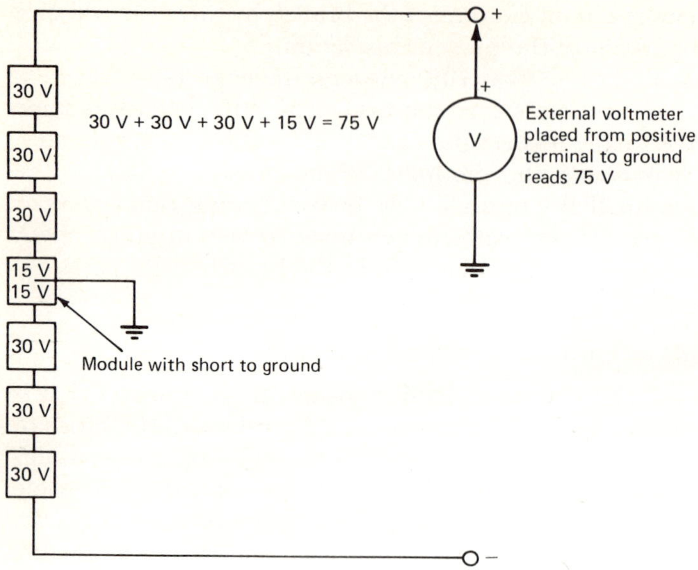

Figure 6-11 **Locating a ground fault with an external voltmeter.**

reading is displayed. If there is a current flow in excess of a few milliamperes, things are not as they should be.

A short or a leakage path to ground may occur in a defective varistor. Checking the varistors involves removing them from the circuit, or more simply disconnecting them from ground. If the leakage to ground has disappeared when the varistors are removed, a replacement part is needed. Tampering with the varistors will bring us in contact with dangerous quantities of electricity; it is recommended that the varistors be removed at night and the leakage test be performed the next sunny day.

A short or leakage path to ground can usually be located by taking a voltage reading from the positive terminal to ground of a floating branch. With the disconnect switch open, the ground connection is severed and the output voltage will show a typical open-circuit reading. A voltage measurement across the output of a floating branch with a short to ground in it will not reveal any sign of trouble. A malfunction will become apparent only when we measure the voltage to ground. Such a measurement must be made with an external voltmeter that has its probes placed on the positive test point and ground (the junction box frame should be grounded and is a likely place to make a ground contact). Since the ground at the junction box and the ground at the fault are quite nearly the same point, we in effect are measuring the voltage from the positive terminal to the fault. This fact is demonstrated in Figure 6-

11. Since we know the typical open-circuit voltage for a module, we can calculate immediately where the fault must be. For instance, if our voltage-to-ground reading is 75 V and our module open-circuit voltage is 30 V, the short to ground must be in the third module from the top of the string. If more than one ground fault occurs in a branch, the one nearest the top of the string will be evident first. Only when this module is replaced or the fault repaired will we notice the next ground fault. When the sunny-day voltage from either branch terminal to ground is zero, all shorts to ground have been eliminated.

A leakage path to ground that is not apparent in the previously described test will be difficult to find; doing so will probably entail testing the modules individually. Access to the back of the modules or test points from each module junction will make it possible quickly to pinpoint where a leakage path exists. Connect the negative lead of the ammeter to ground, and take current measurements at successive modules with the positive probe by starting at the negative terminal and moving up. Notice in Figure 6-12 that the current readings are positive below the faulty short connection and become negative above. The leakage path

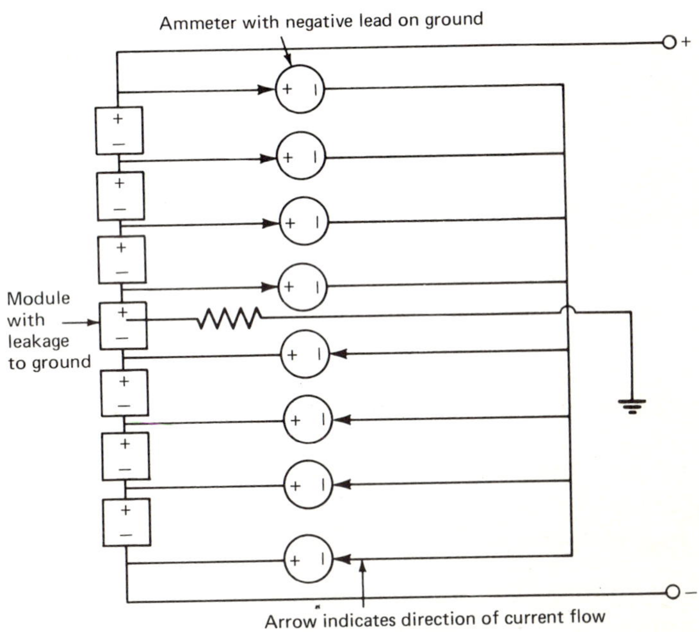

Figure 6-12 **Ammeter with negative lead on ground probes successive module junctions in search of a module with leakage to ground; current polarity changes on either side of the ground fault.**

is somewhere between the two points that registered a change in current polarity. Leakage current in an entire PV array should not exceed 5 milliamperes. Should leakage become as high as 10–20 milliamperes, the shock could be fatal to a human being.

Detecting Open- or Short-Circuited Modules Using a Multimeter

A branch that is connected to the main bus may have an output voltage that is lower than the array voltage and yet have no leakage to ground. A shorted or open-circuited module or a low-current module can all produce this effect on an array. This problem may or may not be apparent when the branch is disconnected from the main bus and the open-circuit voltage is measured. A defective module could perform poorly when loaded down and yet have enough current to appear normal when an open-circuit reading is made. Figures 6-3 and 6-4 demonstrate how both the diode across an open circuit and a short across a module divert current around part of a string and effectively cancel that part's contribution to the branch. When the branch current exceeds the maximum that a low-current module can produce, the diode turns on and shunts current much as an open-circuited module does.

The short-circuit current of each module in a string can be measured if access to the back of the module is possible or if test points are available. Disconnect the branch from the main bus and put a suitably scaled ammeter across each module. A defective module will register little or no current.

Selective shading of modules in a branch that is connected to a load by the main bus may pinpoint a module that has one of the above problems. When the light is cut off from a properly functioning module, the branch output voltage will drop proportionally. However, a defective module contributes little or no power to the output, and cutting off its light will not appreciably affect the output voltage. A two-person team with one person outside shading successive modules and the other person inside watching the branch voltmeter is a simple and safe technique for locating bad modules.

Detecting Faults Using an *I-V* Curve Tracer

An instrument that is capable of safely making a current-versus-voltage curve of the total array can be a valuable tool for discovering array problems. The panel-mounted connectors that are wired to three test points in each branch circuit and the array, two across each current

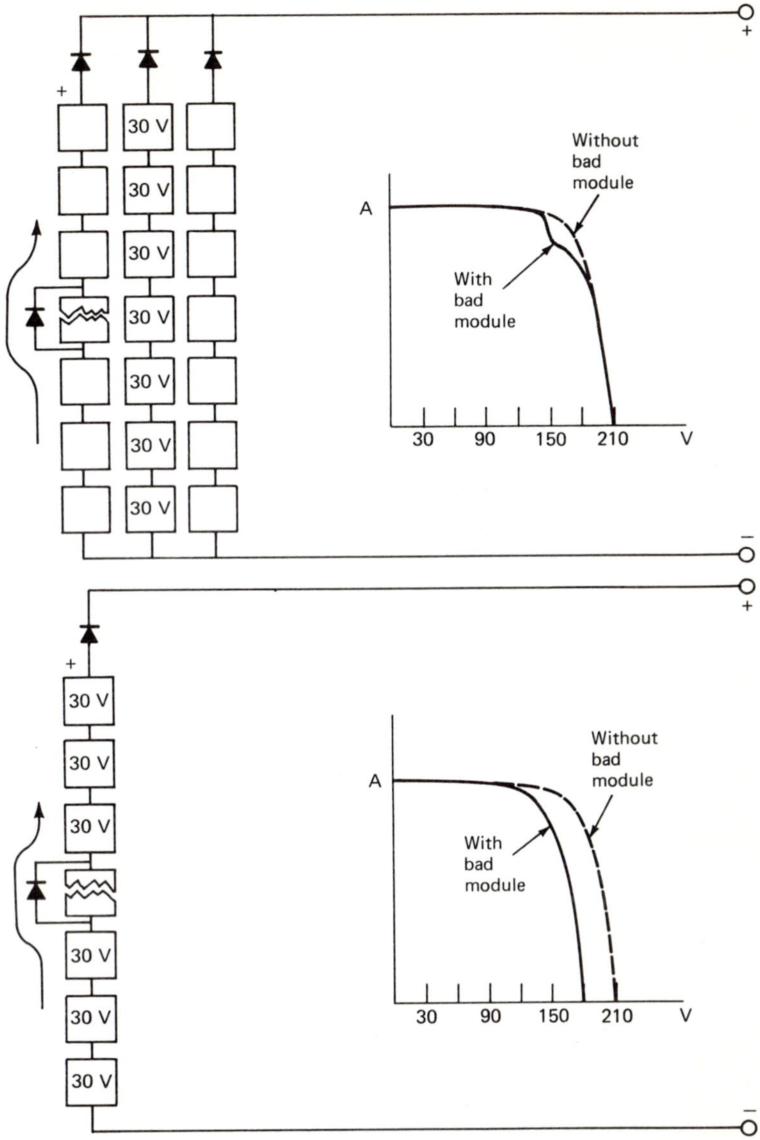

Figure 6-13 Current-versus-voltage curves of an array, and a branch with an open-circuited module in it.

shunt and one on each positive terminal, are available for making such a test.

An initial check involves taking an *I-V* curve of the entire array with modules both in series and in parallel when the array is disconnected from any load, battery, or power-conditioning equipment. Detecting shorts to ground requires that the array be grounded, while open- or short-circuited modules can be discovered whether the array is grounded or not. If there is an open or short circuit someplace in the array, the defect will manifest itself as an indentation in the *I-V* curve. The voltage level at which this indentation occurs will generally indicate where in one of the branches the problem exists. If the abnormality occurs at 70 V, for instance, the problem should be at a point in one of the strings that is 70 V above the negative terminal. The individual strings must be tested to ascertain which one is defective.

Upon discovering a "sick" curve, the next step is to make an *I-V* curve of each string or branch with the negative terminal grounded by keeping the branch connected to the main bus. A healthy branch will exhibit a normal curve with the expected open-circuit voltage and short-circuit current. A defective branch will also have a normally shaped curve, but will drop off to a low open-circuit voltage. If the defect is caused by a short to ground, opening the disconnect switch or eliminating the negative ground connection will restore a normal *I-V* curve trace. Should the low open-circuit voltage persist, the problem must be caused by short- or open-circuited module (with diode protection). Figure 6-13 depicts an array and a branch with a module defect; their respective normal and abnormal *I-V* curves are shown for each condition. By using the selective shading technique and taking successive curves, the malfunctioning module will be found.

The overall troubleshooting strategy is quite simple. First we determine if the array indeed has a problem by taking an *I-V* curve of the entire array and then of each branch. Does the array *I-V* curve have a normal shape and are all the branch *I-V* curves virtually identical? We then make an insolation measurement and compare the short-circuit current of a normal branch with the expected current given the module's specifications and the present insolation. If a branch circuit is not operating properly, we will measure whether there is a short or leakage path to ground in the branch while it is floating by placing a curve tracer from the positive output to ground. If leakage is present, we can find the leakage spot by examining the shape and values of the *I-V* curve. Should leakage not be discovered an open-circuited, shorted, or low-current module is most likely giving us trouble. The defective module can be pinpointed by selectively shading modules in the branch and measuring the output voltage or *I-V* curve.

7

THE PHOTOVOLTAIC ENERGY SYSTEM

Solar electricity performs no function until it is integrated into a system that matches the power supply with its load. To size a photovoltaic generator to meet a particular electricity demand, we must first know the energy requirement and electrical characteristics of the load.

There are a variety of electric-power-conditioning devices that usually must be added to a photovoltaic array to convert the solar energy to a form that can readily be used by a load. Many energy applications require electricity during the nighttime or on cloudy days, and a photovoltaic array is unable to provide this. Common loads often require that the electricity be alternating current (AC), while a photovoltaic cell only produces direct current (DC). Electric loads usually can operate only with a relatively constant-voltage power supply, while a photovoltaic array's voltage can fluctuate quite radically, depending on the amount of sunlight and the size of the load. The photovoltaic array produces maximum power only when it is operating at its maximum-power voltage; a load will often force the array to operate at a voltage that is less or more than optimum. The application of a photovoltaic power source to meet a power demand may require using power-conditioning equipment to address one or more of these three considerations: energy storage, voltage regulation, and DC-to-AC conversion.

THE NEED FOR ENERGY STORAGE

The need for energy storage must be established before the added cost of storage equipment is warranted. Some energy applications function satisfactorily within the time frame of sunlight hours, and therefore can function directly off a photovoltaic array. Other operations require a continuous supply of energy, and therefore storage must be used to transfer photovoltaic electricity to cloudy or nighttime periods. The timing of the load energy demand will determine whether storage is indeed needed.

Exclusively daytime jobs often are associated with the effects of solar heat. A water pump used to irrigate farmland and a fan used to evacuate hot air from a building both serve to cool or moisten something that has been subjected to the sun's heat. In such cases, photovoltaic output will follow the demand; as sunlight and heat increase, the solar cell's capacity to do work also increases.

Tasks that can be performed at any time can conveniently be done during sunny hours. Pumping water into a storage tower, desalinating salt water, freezing water, washing clothes, generating hydrogen gas, and synthesizing fertilizer are example of possible daytime jobs. Some of these direct uses of solar electricity can act as a form of energy storage. Generating hydrogen gas, for instance, produces fuel that later can be burned. Pumping water uphill builds up the potential for harnessing gravitation to do work.

Continuous or sporadic jobs require a reservoir of energy that is available at any time of the day or night. Electric appliances, machines, and lights in a residential, commercial, or industrial facility need energy at any time, and in the modern age people are accustomed to having a surplus that can be tapped whenever it is wanted.

Assessing the need for storage involves comparing the load energy input with a hypothetical photovoltaic array's energy output as they both vary throughout a day and over the entire year. A plot of power versus time of day and energy versus time of year for both the photovoltaic array and the load will demonstrate how well the energy supply and demand are matched.

Photovoltaic Output and Load Input throughout the Day

Sizing a photovoltaic array to meet a load requirement involves considering the total energy and the power level of both the solar-cell array output and the load input. The initial step is calculating the average daily energy for a load under consideration at a particular time of year. We then determine the size of the solar-cell array required to satisfy this demand at that time of year. Next, we assess the power level or timing of the photovoltaic energy output and the load energy input throughout the day. When does the load demand energy and when does the array supply energy? If the two do not coincide, either the load profile will have to be adjusted to accommodate the array's output pattern, or an energy storage system will have to be used to distribute the array's energy over a broader time period.

A load's energy demand and a solar-cell array's energy output are calculated using the techniques outlined in Chapters 1, 3, and 5. We first calculate how much energy in kilowatthours is required by the load on a typical day of a month under consideration. To size a solar-cell array to meet this demand, we position a hypothetical array at the site in question with a south-facing tilt angle equal to the latitude. Once we have determined the solar energy available on the solar-cell surface during one day (in kilowatthours per square meter), we substitute the daily solar energy available, the corrected efficiency, and the packing factor of our hypothetical array into Equation (5-1) and solve the equation for area. Our answer will indicate the approximate number of square meters of solar-cell array needed to satisfy the electricity requirements of the load for a typical day at the site in question.

Comparing the power level of the array output and the load input at different times of the day is facilitated by plotting the load-power/time-of-day curve and the array output-power/time-of-day curve on the same graph. The load-power curve is developed using the technique established at the end of Chapter 1, while the photovoltaic-power curve is estimated by calculating the PV array's peak power and plotting the power/time-of-day curve (see Chapter 3). Looking at the example in Figure 7-1 of a residential load-power input curve and array output plot, we see that the array's time of peak production does not coincide with a typical residence's time of peak power usage. In fact, that is a considerable amount of energy needed during the dark hours that a solar-cell array cannot directly provide.

Bringing the array and the load daily power levels into closer agreement involves one or both of two options: The load can be adjusted to operate only during sunlight hours, or an energy storage system can be used to transfer some of the array's excess power during the day to fill power needs at night. Sunlight can vary quite erratically during the day, with cloud cover able to reduce the power level to near zero. Only certain dedicated loads, such as water pumps, air fans, or electrolysis hydrogen generators, are able to operate under such conditions of changing power.

For larger and diverse loads, such as those found in a residence, the second option is more suitable. Storage systems, however, result in certain energy losses, and this generally would encourage the direct use of

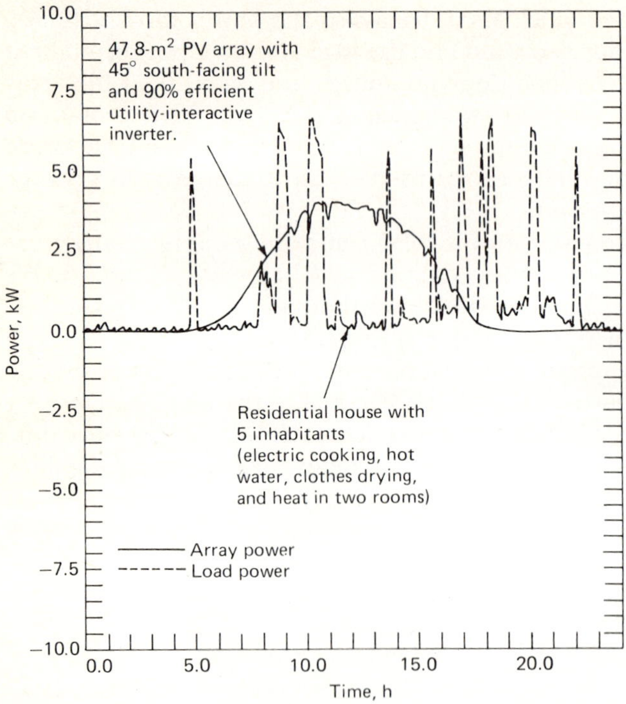

Figure 7-1 **Residential load power input and residential roof array power output versus time of day for a residence and PV array in the Boston area on May 3, 1981. (Northeast Residential Experiment Station, M.I.T. Lincoln Laboratory.)**

as much of the solar power as possible. A combination of adjusting all possible loads to daytime use and storing surplus solar energy for nighttime use will achieve the optimum energy utilization for a combined photovoltaic array and storage system.

Photovoltaic Output and Load Input throughout the Year

Until now, we have been preoccupied with matching a photovoltaic array output with a load input for a particular day of the year. However, a load's energy needs and the energy received from the sun vary, and so our match during one month would be considerably altered at a different time of the year. We must compare how the load energy use and the solar energy availability fluctuate vis-a-vis each other as the year progresses. Plotting load energy versus time of year and photovoltaic array energy versus time of year on the same graph, as shwn in Figure 7-2, provides a simple and quick comparison. An intersection of the two curves represents a date when the array output and load input are well coordinated.

The annual load energy input and PV-array energy output profiles will often be poorly matched. Under such circumstances we are faced with a trade-off; if the array is designed to meet the load's maximum need, we may end up having a large surplus during other times of the

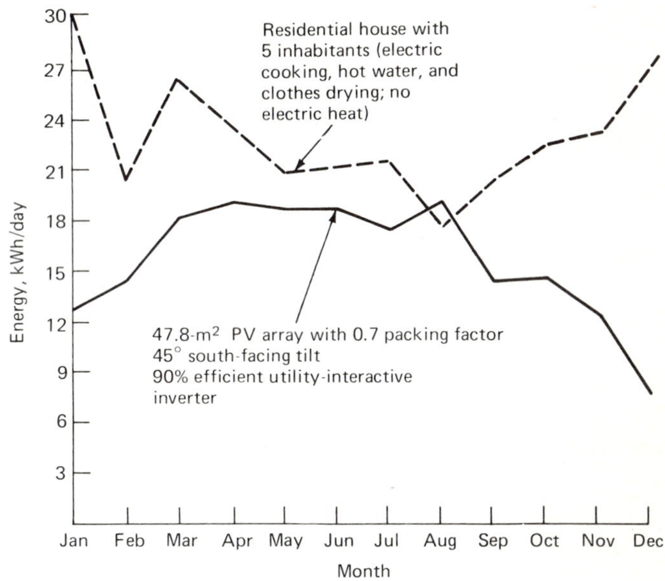

	PV energy output, kWh/d	Load energy input, kWh/d		PV energy output, kWh/d	Load energy input, kWh/d
Jan	12.4	30.0	Jul	17.6	21.6
Feb	14.2	20.3	Aug	18.9	17.7
Mar	18.0	26.2	Sep	14.2	20.5
Apr	18.8	23.2	Oct	14.6	22.3
May	18.3	20.9	Nov	12.3	23.1
Jun	18.4	18.4	Dec	7.6	27.3

Figure 7-2 Residential load energy input and residential roof array energy output versus time of year for a residence and PV array in the Boston area, measured between May 1981 and April 1982. (Northeast Residential Experiment Station, M.I.T. Lincoln Laboratory.)

year, while if we design the array for the load's minimum need, we could have an energy deficit for months. There are various options available for resolving this dilemma.

First, the load can be adjusted to follow the sun's annual energy profile at the site. The photovoltaic array can be sized to meet the load's maximum energy need, and during times of surplus, excess energy can be diverted by switching in auxiliary loads or power dumps.

Second, an array can be sized to meet the average energy need of the load throughout the year. In this way, energy surpluses and deficits can be minimized. Storage and/or an additional power source will be needed to provide energy during times of insufficient sunlight.

Third, the tilt angle of the photovoltaic array can be altered so that the power output of the array is maximized at a certain time of year rather than over the entire year. If the load in winter is greater than that in summer, the array can be angled to maximize the winter sunlight. A surface's optimum tilt angle over an entire year equals the latitude angle. The optimum tilt angle at any one time of year equals the latitude angle minus the declination angle. (The diffuse and reflected components of sunlight may alter the optimum angle by about 1% to 15%.) Figure 3-18*b* shows how changes in surface tilt angle affect the array output throughout the year. Employing this option could result in the array producing less than the total annual maximum energy possible, and therefore not realizing its full potential. A variation on this option would involve changing the array tilt two to four times a year to keep its surface more perpendicular to the sun's rays. Most arrays, however, are mounted in one fixed position. Changing the tilt angle will rarely cause the array output power level to perfectly match the load profile; consequently, a storage or backup power supply will still be needed.

Electricity storage will generally be needed to transfer surplus energy produced during sunny days to energy-deficient periods, night or less productive times of the year. Often it will not be economically feasible to design a photovoltaic system with enough storage to meet the longest recorded period of poor sunshine, since this would result in a system that is oversized for most other times of the year. In sizing the system, a balance must be struck between keeping the array cost to a minimum and having sufficient output capacity to meet the load's usual energy demand. This trade-off often results in a PV-storage system that satisfies the load's energy requirements most but not all of the time. Consequently, some forms of backup power may need to be available for those few times of the year when because of weather or load conditions the PV system cannot meet its obligations. Backup power is generally provided by the utility or a diesel generator. If no backup power is available, brief periods of little or no power may be preferable to an oversized and expensive array. On the other hand, small systems in remote locations such as those used to power such loads as beacons, radio repeater stations, and railroad crossings, routinely have oversized photovoltaic arrays because no backup power is available and power outages cannot be tolerated. In such cases, the increased cost of a large PV array and discarded energy is deemed worthwhile.

Electricity Storage

There are two common ways of storing electric energy: the battery and the utility grid. The battery stores electricity in a chemical state, while the utility grid acts as a "bank" or repository for surplus energy that will be withdrawn at a later time. A battery can store energy efficiently for only a short period of time (a few months or less). It is a good tool for transferring sunny-day photovoltaic energy to nighttime or cloudy-day use, but it will not be able to transfer energy over long periods, such as storing excess summertime solar energy to meet a higher wintertime energy demand. Buying and selling power to the utility grid is a better method of long-term energy transfer (if the utility will permit it). When the relative costs of a suitable-sized battery are compared with those of a utility-interactive inverter, grid storage will generally be more economical than battery storage for sites that have ready access to the utility power network (given 1983 prices).

Utility Grid Storage

The utility grid could be used by a photovoltaic array to store energy on both a short-term and a long-term basis. Excess photovoltaic energy generated during the day is sold to the utility, while the lack of photovoltaic energy during the night and on cloudy days is supplied by buying power from the utility. Similarly, excess energy generated during any time of the year (typically the summer) can be sold to the utility, while a deficit of energy available from the photovoltaic system at another time of year (typically the winter) can be supplemented by buying energy back from the utility. The photovoltaic array can be sized so that the total annual energy output roughly equals the total annual energy demand, and thus the net energy obtained from the utility is near zero.

A recent federal law (PURPA, the Public Utilities Regulatory Policy Act) requires utilities to buy power from cogenerators, such as utility-connected photovoltaic systems, at a reasonable rate, which probably will be between 50% and 100% of the utility's selling price. At a 50% buyback rate, a zero net energy flow between a solar electric system and the utility will not mean a zero net energy cost. From the point of view of the residence, the energy sold to the utility will be worth half as much as the energy bought from the utility. Different arrangements are being developed between utility companies and the cogenerators attached to their grid. The two-ratcheted kilowatthour-meter approach is used when the buyback ratio is other than one; energy flow from the PV power source to the utility is logged on one meter, and utility power to the load is logged on another. The one bidirectional kilowatthour-meter approach is used if the buyback ratio is one, since this meter operates both forward and backward. The billing arrangement may allow energy credit to transfer from one month to the next, or may require that energy surpluses be used within a particular month.

A utility-interactive photovoltaic system is made up of various components that enable a proper and safe interface between a photovoltaic array and the utility grid. The direct current produced by a photovoltaic array is converted to the alternating current produced by the utility with a device called a *power inverter.* (More details about the inverter are provided in a later section.) After DC-to-AC conversion has been performed, the photovoltaic power must be connected to a point that is electrically common to both the utility and the local loads, as shown in Figure 7-3. The photovoltaic connection is made on the utility side of the main circuit breakers so that all loads can be disconnected during a malfunction. One or two energy meters (kilowatthour meters) will need to be installed in the line between the inverter output and the utility connection to register the energy flow in either direction for billing purposes. Finally, the inverter output and load input are connected to the utility distribution transformer.

The interface between a residential- or commercial-scale photovoltaic cogenerator and the utility will normally be performed through the utility's pole transformer. The utility's AC electricity is distributed from the power plant at a very high voltage to keep the power I^2R losses to a minimum (there is a greater power loss in the transmission lines at high currents than at high voltages: $P = I^2R = V^2/R$). This high voltage from the power plant is stepped down in two or more stages at substations and at pole transformers. For example, a power plant may transmit electricity at 250,000 to 48,000 V AC; this voltage is stepped down to 23,000 and then to 2300 V. The substation voltage is finally reduced to the 210 to 240 V available at most establishments by a pole transformer near each site. The pole transformer output winding has a center tap that is grounded in the middle to produce a 110-V drop from either output to ground, as shown in Figure 7-3. Small to medium-sized residential or commercial loads are connected across one of the two 120-V outputs to ground, while more substantial loads (an electric clothes dryer or a central space heating or air conditioning system) are connected

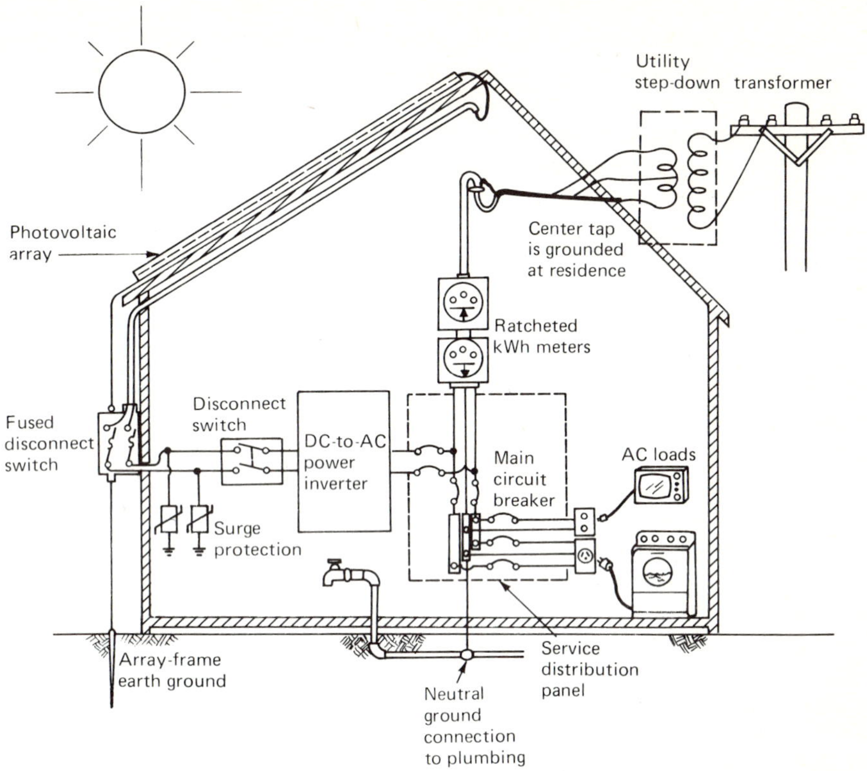

Figure 7-3 **Block diagram of a utility-interactive photovoltaic system. (M.I.T. Lincoln Laboratory.)**

across the entire 230-V transformer output. A medium- to small-scale utility-interactive photovoltaic system will be connected across the transformer's 230-V output, and any power sold to the utility will be sent through this link.

There are two safety features that must be added to a utility-interactive circuit to protect both people and equipment from damage: a loss of utility power disconnect switch and electrical isolation devices between the utility and PV power. In addition, there are electrical problems, involving AC power factor and harmonic distortion, associated with converting DC to AC and interfacing with the utility. These electrical issues will be discussed in a later section on DC-to-AC conversion circuit design.

In the event of a utility blackout, repair personnel will need to fix the utility's circuit without any power on the lines. A utility-interactive inverter could send power into the lines and therefore must be disconnected whenever a utility shutdown occurs. This is accomplished automatically with a voltage-sensing circuit and a relay between the photovoltaic array and the utility grid.

There is a real danger that utility power might feed back into the photovoltaic power system or that DC electricity might be injected into the utility grid. The photovoltaic array must be electrically isolated or somehow protected from the utility. A transformer is composed of two separate windings that transmit alternating current through a magnetic field at their core. There is a limit to how much AC a transformer can transmit, and DC is unable to pass through a transformer. A transformer would serve well to isolate the photovoltaic array and the utility.

A transformer that can handle a large amount of power can be expensive and will reduce the inverter's overall efficiency. Alternative designs, therefore, are being suggested that are capable of sensing excessive alternating or direct current and disconnecting the PV system from the utility. Fuses can prevent an inrush of current from the utility into a faulty inverter, and a DC-sensing circuit between the inverter and the utility can detect DC injection and trip a relay or contactor that disconnects one from the other.

Using an isolation transformer will permit grounding of one of the photovoltaic array terminals. A utility-interactive design without a transformer will not function with a grounded array because the utility's distribution transformer has its center tap grounded: a ground connection on either side of the inverter will short out half of its cycle. Nonetheless, in a nonisolated system there always is a "virtual" ground connection at the utility transformer that will prevent the array voltage from floating high as a result of atmospheric static (as discussed in Chapter 6).

Simple utility-interactive photovoltaic systems have been developed that demonstrate reliable and efficient performance (examples are found in Appendix A, State of the Art). The power-conditioning components shown in Figure 7-4 are typical of the many utility-interactive photovoltaic systems that have been established by M.I.T. Lincoln Laboratory under the auspices of the Department of Energy. As is evident in this figure, utility-interactive line-commutating inverters are being marketed that are compact (about 5 ft^3 or $^3/_4$ m^3 for an 8-kW unit) and have a relatively low parts count and cost. The major obstacles to the wide use of these systems are the cost of solar cells and the technical and political logistics of the utility interface. A small number of successful PV cogenerators are already interconnected with utilities; technical and political problems may, however, arise when large numbers of small-scale utility-interactive systems are considered.

(a)

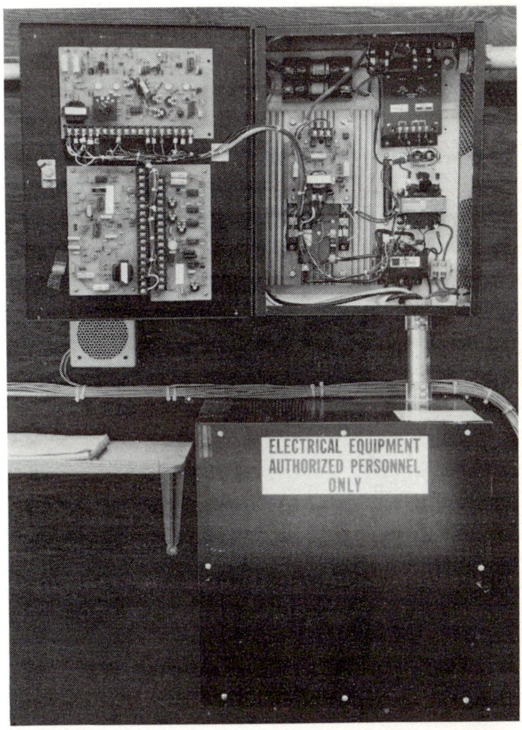

(b)

Figure 7-4 **Power-conditioning and monitoring components of a utility-interactive photovoltaic system at NERES: (*a*) a complete system with monitoring and display junction boxes; (*b*) internal circuitry of a utility-interactive inverter, the box labeled *Gemini* at center of photo (*a*). (Photos by M.I.T. Lincoln Laboratory.)**

Battery Storage

A storage battery can be used to store photovoltaic energy on a short-term basis. The type of battery that can easily store solar electricity has an efficiency (energy retrieved divided by energy deposited) that decreases with storage time. Storage batteries, therefore, rarely are used to store energy for longer than a few months. Batteries cannot be used for long-term storage of photovoltaic electricity, such as might be required if excess energy generated during summer days needed to be saved for low-solar-power periods during the winter. Battery storage, however, does lend itself to storing sunny-day energy output for night or cloudy-day use. Battery storage is not presently cost-effective for loads that have easy access to the electric utility grid; consequently, battery storage typically is used for stand-alone and remote site applications.

A battery stores electric energy in a chemical state of matter. The basic component of a battery is the *cell.* Much like solar cells, battery cells are modular and can be viewed as electric building blocks; placing cells in series increases the voltage output, while putting cells in parallel increases the current output. The battery cell is a specifically engineered sandwich of two or three special materials juxtaposed in a manner that promotes or facilitates the transfer of electrons from one material to the other. When one of two substances has acquired more electric charge than another, an electromotive force or voltage builds up across the two materials. This voltage drop across the two differently charged *plates* is made available at the output electrodes of the battery and provides the impetus for driving electric current through a circuit.

Certain types of cells (called *primary cells*) automatically convert chemical energy into electric energy when they are first constructed; after the battery has been discharged, this conversion will not recur. A primary cell cannot be effectively recharged unless additional chemicals are added to the cell (as is done with fuel cells). Carbon/zinc flashlight batteries are primary cells and thus are discarded once they have been discharged.

A second type of cell (called the *secondary* or *storage cell*) is capable of repeatedly exchanging electric energy and chemical energy by going through many charge and discharge cycles. Lead-acid and nickel-cadmium batteries are two common types of secondary cells, with the former being more attractive for photovoltaic-energy storage because of its lower cost per unit of storage capacity. A comparison of the characteristics of lead-acid and nickel-cadmium batteries is found in Table 7-1, and an explanation of the various terms follows.

There are five basic battery characteristics and concerns that must be understood in order to use a battery properly with a photovoltaic array: storage capacity, storage efficiency, state of charge, operation procedures, and maintenance. The amount of electric energy that a particular

Table 7-1 A Comparison of the Features of the Nickel-Cadmium and the Lead-Acid Storage Battery*

	Lead-acid	*Nickel-cadmium*
Type	Medium-rate, deep discharge, lead calcium grid	Medium-rate, cycle service, vented pocket plate
Rated capacity at 77°F, 8 h, Ah/cell	100–900	10–400
Nominal discharge cutoff voltage, V/cell	1.75	1.0
Nominal voltage, V/cell	2.45, varies with SOC	1.25, fairly constant with SOC
Available capacity vs. temperature, % of rated	70% at 32°F 20% at −20°F	90% at 32°F 65% at −20°F
Nominal energy efficiency, %	70–80	60–70
Nominal cycle life for 80% discharge, cycles	1000–1500	1500–2000
Nominal calendar life without cycling, years	10–20	24
Energy density, Wh/lb	6–13	9–10
Required maintenance	Water replacement Charge equalization Protection against freezing and temperature extremes	Water replacement Occasional full discharge
Charge control	Sensitive to overcharging	Can accept 5–10% overcharging
Hazards	Electric shocks, hydrogen gas evolution, acid leaks/spills	Electric shocks
Cost,† $/kWh in 1979	165–250	300–1000

Source: Y. P. Gupta and S. K. Young, *Design Handbook for Photovoltaic Power Systems,* Science Applications, McLean, Va., 1981, p. 80.

*Currently available battery options for grid-connected photovoltaic systems (with frequent battery cycling)

†For comparison only; actual prices fluctuate widely.

battery is capable of storing must be ascertained. A battery's efficiency, or the percentage of stored energy that can actually be retrieved from it, is an important characteristic that will enter many of our calculations. There must be some way of determining a battery's state of charge, or how much of the battery's storage capacity is still available (a state-of-charge meter is similar to a fuel gauge in a car). Proper operating and maintenance procedures need to be observed when using batteries in order to prevent damage and ensure maximum life.

A lead-acid battery's storage capacity can be rated in either ampere-hours or watthours. The ampere-hour specification indicates how much current can be obtained from a battery, irrespective of the rated voltage. The watthour capacity is a more valuable piece of information, since it denotes the amount of energy a battery can store. (Manufacturers tend to rate their batteries in ampere-hours; the designer will need to convert the ampere-hour specification into a watthour figure.) A battery usually is specified by a maximum discharge percentage that indicates the deepest discharge that a battery normally can tolerate. If a battery has a "60% allowable depth of discharge," 40% of the energy stored in the battery must remain and should never be removed.

The ampere-hour rating of a battery is valid only at a specific discharge rate, which usually ranges between 5 and 30 hours, and at temperatures within approximately $\pm 20°C$ of room temperature. The ampere-hour figure indicates how many amperes can be expected to flow from the battery during the specified discharge period. For instance, a 100-Ah battery with a 20-h rating will provide 5 A continuously for 20 h (5 A $\times$ 20 h = 100 Ah). If the battery is discharged with a current in excess of 5 A, the ampere-hour figure will not be met, while if the discharge rate is less than 5 A, the ampere-hour value will be exceeded. The curves in Figure 7-5 demonstrate this principle.

The watthour storage capacity can be determined by multiplying the battery's ampere-hour rating by its average discharge voltage (at about 60% state of charge) and by the usable percentage of the battery, as in Equation (7-1).

$$\mathrm{WC} = \mathrm{AH} \times V \times \mathrm{BF} \tag{7-1}$$

where WC = usable watthour capacity, Wh

AH = ampere-hour capacity, Ah

V = average battery voltage at 60% SOC and room temperature, V

BF = fraction of battery that is available for use, 0 to 1

For instance, if a 24-cell battery has an average voltage of 48 V, an ampere-hour rating of 200 Ah, and a battery fraction of 0.8, the watthour

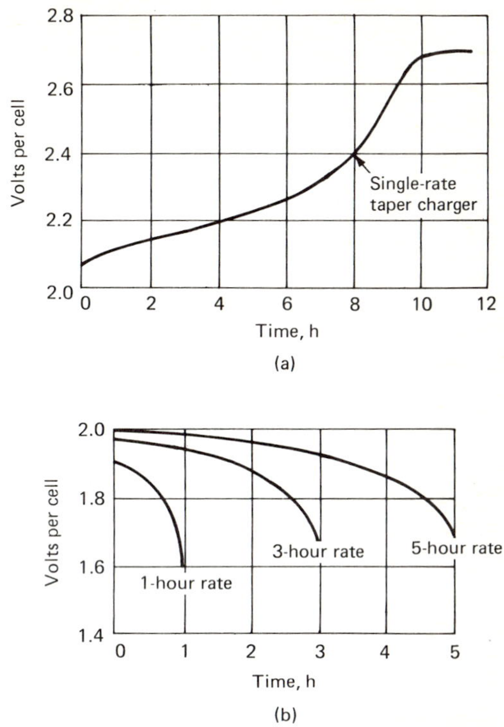

Figure 7-5 A lead-acid battery's voltage versus (*a*) charge and (*b*) discharge time. (G. Smith, *Storage Batteries,* Isaac Pitman & Sons, London 1964, p. 43.)

storage capacity will equal 7680 Wh. Once the battery is fully charged, this quantity of energy can be removed only if the discharge rate is no higher than specified and if the battery temperature is not significantly lower than room temperature.

A battery's storage capacity degrades with age at a rate that largely depends on the number of charge/discharge cycles, the depth of discharge, and the average operating temperature. Figure 7-6 shows that as the depth of discharge and the mean battery temperature increase, the number of useful charge/discharge or life cycles decreases. Batteries are typically said to have ended their useful life when the storage capacity has dropped below 80% of the nominal capacity. A storage battery can be expected to operate for 5 to 25 years, depending on the above-mentioned conditions and its specific design (assuming that proper operation and maintenance procedures are followed).

There are two ways of measuring a battery's efficiency: watthour and

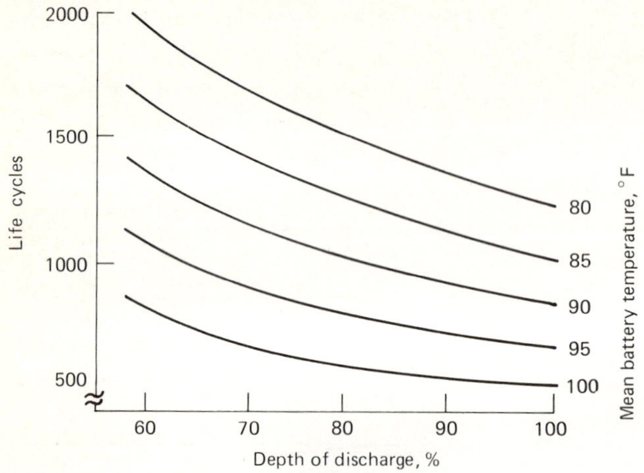

Figure 7-6 The effect of depth of discharge and temperature on lead-acid battery life. (Bechtel National, Inc., *Handbook for Battery Energy Storage Photovoltaic Power Systems,* Department of Energy Report SAN-292-T1, 1979).

ampere-hour. The watthour efficiency is the battery's maximum energy output divided by the maximum energy input. The ampere-hour efficiency is the battery's total current output divided by the total current input for a full charge/discharge cycle. Since the charging voltage is appreciably higher than the discharge voltage, the ampere-hour efficiency figure will be somewhat misleading and will always be higher than the watthour efficiency under the same conditions. The current into the battery will nearly equal the current out (there is almost nowhere else for the electrons to go, other than to slowly self-discharge within the battery). The voltage-in, however, is significantly higher than the voltage-out, resulting in an appreciable difference between power-in and power-out. A new lead-acid battery's ampere-hour efficiency typically is around 95%, while the usual watthour efficiency is about 85% (using a new battery at an ambient temperature and a specified discharge rate).

Battery temperature and storage time have similar effects on both storage capacity and efficiency. A decrease in temperature or an increase in the length of time a particular quantity of energy is stored will cause the battery's performance to suffer. Battery efficiency and capacity decrease with time because of a self-discharging current within the battery, but drop off significantly only after a month or more. A decrease in a battery's temperature below room temperature produces a conditional decrease in efficiency and storage capacity. At lower temperatures, the battery's discharge voltage is decreased, and therefore power output is reduced. When the temperature returns to a normal range, efficiency and storage are restored to their original values. A lead-acid battery's storage capacity decreases about 1% for every drop of 1°C in the battery temperature. Extreme temperatures, whether very cold or very hot, will permanently damage a battery and clearly must be avoided.

It is important to know the level of energy or state of charge that exists in a battery at any point in time; we thereby can tell how much energy is available. A battery's state of charge (SOC) is monitored periodically or continuously using one of several techniques. The voltage of a lead-acid storage cell fluctuates quite linearly with state of charge and therefore is a good indicator. If the charge across the two plates of a lead-acid cell is at its maximum, the open-circuit voltage is about 2.3 V. When the state of charge is at its minimum, the voltage drops to about 1.7 V. A graph of cell state of charge versus cell voltage is shown in Figure 7-7. (Note how the voltage of a lead-acid battery varies quite linearly with changes in SOC between about 1.7 and 2.3 V per cell.) Other methods of determining SOC involve measuring the specific gravity of the acidic electrolyte fluid (a high specific gravity indicates full charge, and vice versa), and placing a two-way current-measuring electronic circuit at the battery output to log the net energy flow between the battery and its charger or load.

Proper battery operation requires voltage-regulating protection circuitry to prevent overcharging or excessive discharging. Permanent damage can be inflicted on a battery if it is charged too fast and too

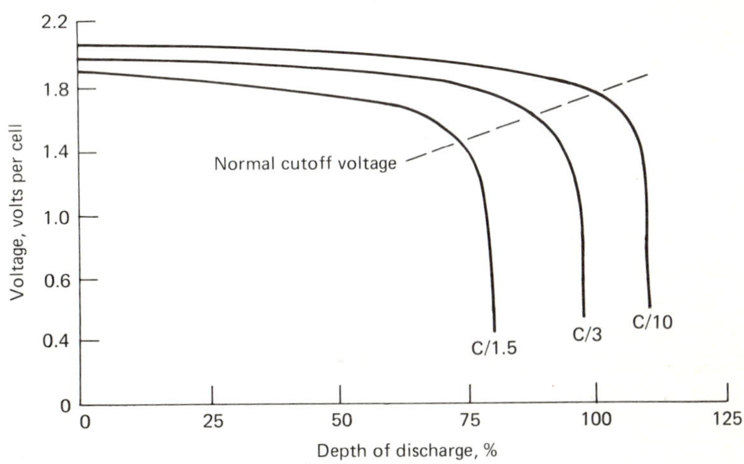

Figure 7-7 **Lead-acid battery voltage versus state of charge; *C/x* is the discharge rate, where *C* = total Ah capacity and *x* = discharge current. (Bechtel National, Inc, *Handbook for Battery Energy Storage in Photovoltaic Power Systems Final Report*, Sandia Laboratory, 1979, p. 3–11).**

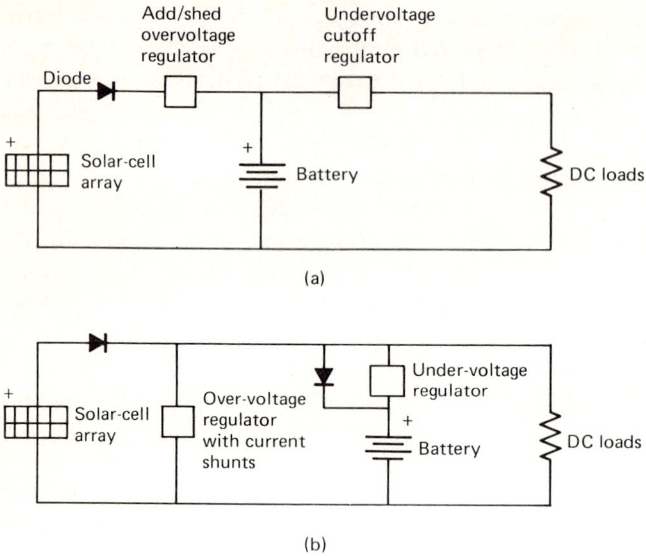

Figure 7-8 **Photovoltaic-system designs with and without battery storage: (*a*) battery storage with add/shed overvoltage regulator; (*b*) battery storage with current-shunting overvoltage regulator; (*c*) battery storage with utility-powered battery charger backup power; (*d*) battery storage with one-way power-flow utility backup power and a DC-to-AC inverter; (*e*) no battery storage with a two-way power-flow utility-interactive inverter.**

long. Low-level trickle charging can be performed indefinitely, since it only offsets the self-discharging current of the battery. However, forcing higher charging currents into a battery when it is fully charged will cause the battery to *gas* (boil off hydrogen and oxygen from the water in the electrolyte fluid). Gassing decreases the fluid level of the electrolyte and thereby exposes and damages the battery's plates. Excessive discharging will cause the plates to disintegrate and also must be avoided. Since excessive charging causes the battery voltage to rise prohibitively high, and conversely excessive discharging pulls the battery voltage too low, we can protect the battery by using one of many different types of voltage-regulating circuits that maintain the battery voltage within an acceptable range or "window." The graph in Figure 7-5 depicts the typical charge/discharge characteristic of a deep-discharge lead-acid bat-

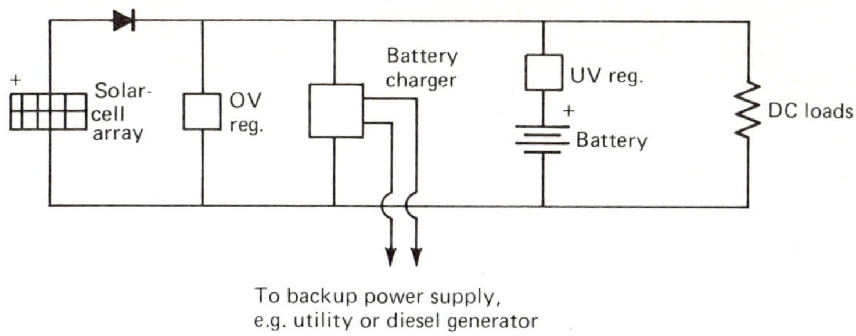

(c)

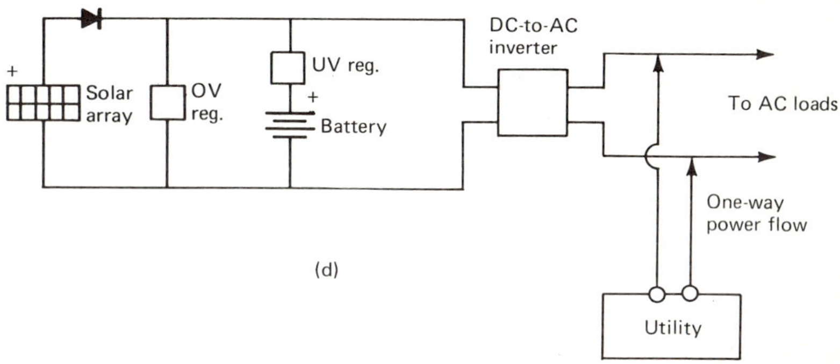

(d)

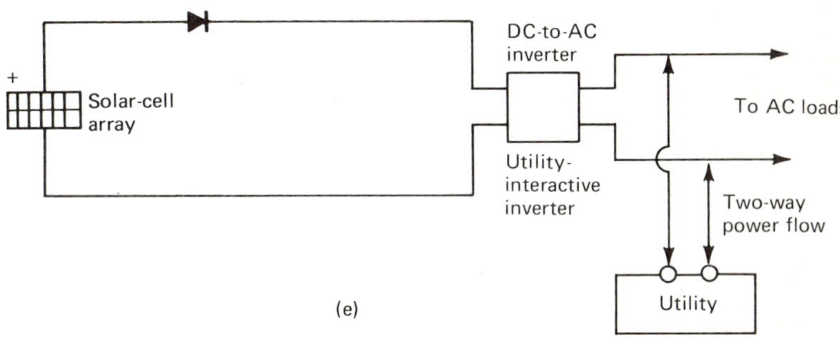

(e)

tery. Notice that the charging voltage is always higher than the discharging voltage, and that the overall operating voltage is kept within a certain range.

Battery maintenance is essential to ensure the longest possible battery life; the two basic tasks are checking the fluid level and battery equalization. Most lead-acid batteries are composed of a liquid electrolyte that diminishes as a result of evaporation and gassing. The fluid level must not be allowed to drop to a level that exposes the battery plates to air. Consequently, the fluid level in each cell must be checked and periodically refilled or "topped off" with distilled water. Under normal operation, some cells in a battery will tend to discharge more quickly than others and, conversely, will not reach the uniform level when charged. Battery equalization involves a carefully monitored battery overcharge cycle that will bring these weaker cells up to the normal capacity. Failure to equalize will cause these feeble cells to further degrade and thereby decrease the battery's efficiency and storage capacity. A battery equalization cycle usually lasts about 10 to 25 hours and is performed once every 1 to 6 months. Equalization with a photovoltaic array can be performed during brief but frequent periods around noon of each sunny day, if the array and battery protection circuit are suitably sized.

There are certain safety precautions that must be strictly observed when using lead-acid batteries. The hydrogen and oxygen gas generated during gassing is highly explosive and must be evacuated and diluted with some type of ventilation system. Hydrogen detection alarms are also recommended. Recombiners have recently been developed that can be used to combine the two gases into their original compound, water. The water is then returned to the battery's electrolyte, eliminating the need to periodically add water. Battery acid is highly corrosive, and steps must be taken to prevent this acid from spilling.

Common system designs for a combined photovoltaic array with battery storage are shown in Figure 7-8*a* to *d*. The basic components are the array, the battery, the voltage- and/or current-regulating battery-protection circuitry, and the load. Figure 7-8*a* shows the battery-protecting regulator devices in series with the solar-cell array and the load. When the battery voltage becomes dangerously high, the overvoltage regulator sheds part or all of the array to reduce or stop the charging current. Conversely, when the battery voltage and charge drop to a relatively low level, this regulator adds part or all of the array to the system circuit. The undervoltage regulator disconnects the load when the battery charge and voltage are below the minimum acceptable level. In the event of a load cutoff, the array is also disconnected from the load and is only allowed to charge the battery. When the battery voltage returns to a permissible level, the undervoltage regulator reconnects the load.

The photovoltaic system with battery storage shown in Figure 7-8*b* has the battery-protection circuitry arranged in such a way that there is a permanent and direct connection between the array and the load. The undervoltage cutoff regulator is placed in the battery leg of the circuit. When a low battery voltage causes the regulator to disconnect the load from the battery, the diode wired across the regulator will nonetheless allow the array to charge the battery. The overvoltage regulator shunts current away from the battery when its voltage rises too high. A variable current shunt can be created using a bank of resistors that are connected in parallel with the array via relays, or it can employ a linear integrated circuit device that biases a transistor, which in turn directs current through one or more resistors.

In both these battery-charging designs, excess array power is either disconnected or dumped into a bank of resistors. Large amounts of solar electricity can be squandered if such a scheme is used. A more imaginative design might put this extra energy to good use by switching in some intermediate loads.

In order to make the photovoltaic system compatible with the utility and most common loads, the photovoltaic DC electricity will need to be converted into AC. Figure 7-8*d* demonstrates how a photovoltaic array, battery, and stand-alone inverter are interconnected. In comparison, Figure 7-8*e* shows a utility-interactive photovoltaic system without battery storage. In the first case, the inverter is capable of standing alone, and requires utility backup power only when the battery storage has been depleted during extended cloudy conditions or a heavy electricity de-

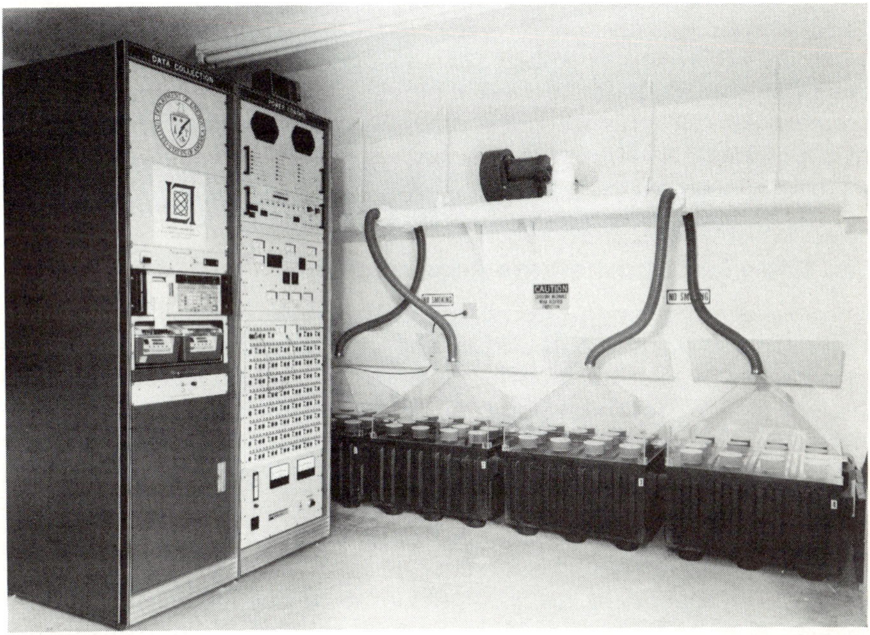

Figure 7-9 **A bank of lead-acid batteries with 40-kWh storage capacity (air ducts are to evacuate dangerous hydrogen gas) and with photovoltaic power controller and data collection system. (Photo by M.I.T. Lincoln Laboratory.)**

mand. In the latter case, the utility grid is the only means of storage, and hence we have a two-way power flow between the utility and the photovoltaic system.

Photovoltaic systems with battery storage are typically used for remote stand-alone sites without access to the utility grid. The current price and maintenance requirements of lead-acid batteries restrict the value of battery storage systems in areas where utility power is readily available. It is hoped and anticipated that more durable and less expensive deep-discharge batteries will be developed. An example of a 40-kWh battery bank used in conjunction with a 15-kW peak photovoltaic array to power a daytime AM radio station is shown in Figure 7-9. The system controller, also shown in this figure, uses an array-shedding scheme to protect the battery from overcharging. The air-venting pipes are necessary to evacuate the dangerous hydrogen gas that is emitted when the batteries are gassing.

VOLTAGE REGULATION

The size and stability of the system operating voltage are critical for both the photovoltaic array and the load. Electric loads generally operate best when the power-supply voltage remains fairly constant, especially when two or more loads are connected in parallel. Maximum power is gleaned from a solar-cell array when it is operating at its maximum-power voltage. Designing a photovoltaic energy system involves integrating both of these concerns.

The Need for Load-Voltage Regulation

Load-voltage regulation often is essential for two reasons. First, many types of loads operate only within a specific range of voltages and will be damaged or malfunction when exposed to voltage levels that are too high or too low. Second, when two or more independent loads are connected to a common power supply, they will interact if the voltage or current is not regulated. Interaction between loads is unacceptable when the goal is any kind of reliable or consistent operation.

An electric load usually fits into one of three categories, resistive, capacitive, or inductive. The current through a resistive load varies quite linearly with changes in voltage, as demonstrated by the load line shown in Figure 4-12. Typical loads that exhibit this basic characteristic are incandescent lights and the heating elements found in electric heaters or stoves. The current through an inductive or capacitive load does not vary linearly with changes in voltage. The current-versus-voltage curve can vary considerably depending on how the inductive or capacitive load is designed (a more detailed description of inductance and capacitance is provided in a later section). The most common inductive load is the *motor*. A wide variety of appliances are equipped with motors, including washing machines, fans, air conditioners, electric clocks, refrigerators, blenders, and record players. There are few capacitive loads in common use; one of the only examples is the fluorescent light bulb.

When the voltage across a load varies, certain possibly undesirable effects result. If the voltage across a resistive load, such as a light bulb or heater, is increased, there is an exponential increase in the power output; the light becomes brighter or the heater becomes hotter. If the voltage across an inductive load, such as a motor, is increased, the rotation speed is proportionally increased. A motor requires more power to start rotating than it does to maintain its rotation; this higher burst of power can be produced by a momentary increase in current. There is a minimum voltage below which a motor and its related appliances will not start. Other loads, such as a radio or television, also require a minimum voltage in order to operate and will fail to function below this cutoff level. All loads have an upper voltage limit beyond which the device will be damaged. A load therefore operates within a specified range of voltages that is rated by the manufacturer. Maintaining a stable voltage within the load's rated operating range will ensure continuous and consistent operation.

When power is initially applied to an inductive, capacitive, or resistive load, there often is an initial surge current that flows before the load stabilizes. Some resistive and capacitive loads, such as incandescent or fluorescent lights, are known to draw 5 to 10 times their normal operating current for a small fraction of a second before leveling off. An inductive load such as a motor also draws between 5 and 10 times its normal operating current when power is first applied, but it does so for a relatively long period of time (a large fraction of a second or more).

The size and long time period of inductive surge currents poses a problem for most power supplies, including a photovoltaic array. When a motor is initially placed across the terminals of an illuminated photovoltaic array, the load's input acts momentarily like a short circuit and therefore draws the solar-cell short-circuit current. Near short-circuit current, the solar-cell array is producing negligible power, since it is not operating close to its maximum-power point, and the output voltage is almost zero. If the current through the load is not large enough to start the motor, the array and motor operating voltage will remain stuck near zero, even though the motor, once in motion, might be capable of operating on the available photovoltaic maximum power. The inductive load needs a regulated voltage to function; even a properly sized photovoltaic array on its own is unable to provide a regulated voltage under low levels of insolation. This problem becomes especially evident when many motor-driven loads are operated directly off a photovoltaic array.

More than one load will usually be attached to a photovoltaic power source at any one time. To prevent interaction between different loads connected to the same power supply, the power source must regulate its output so that it produces a constant voltage or current; the unfixed variable is allowed to fluctuate according to the needs of the various loads that are connected in parallel. The utility power we are accustomed to using is based on a system design that fixes the voltage and allows the current to be determined by the load energy demand. Most electric loads are designed to operate with such a voltage source.

According to Ohm's law ($V = IR$), if the voltage is constant, the current will increase as the resistance decreases. If the voltage is not kept constant, as more loads are attached to the power source, the voltage and current will both vary to accommodate the new load. This change in voltage will alter the power available at all the loads, and hence an undesirable situation of interacting loads will be created.

The Need for Solar-Cell Voltage Regulation

A solar-cell power source poses certain problems in that its output voltage or current does not remain constant or regulated when sunlight and load energy demand change. Examining the current-versus-voltage curve of a solar cell (as seen in Figure 4-6), we notice that the current remains fairly constant during most of its performance (at any one given insolation). This characteristic might be put to good use to prevent load interaction were it not for a few limiting factors: most loads rely on a constant voltage, the maximum-power point of a solar cell is on the poorly regulated knee-shaped portion of the *I-V* curve, and a solar cell's current in fact does not remain constant, since it varies with insolation.

The solar-energy-to-electric-energy conversion efficiency naturally decreases as the solar-cell operating voltage migrates away from the maximum-power voltage. When we state that a solar cell's conversion efficiency is, let us say, 10%, we are assuming that it is operating under optimum conditions, that is, that the attached load or power-conditioning equipment is causing the solar-cell array to operate at its maximum-power voltage. In order to maintain the highest possible energy conversion efficiency and therefore to optimize the economic value of a photovoltaic energy system, the solar-cell array should operate as close as possible to its maximum-power voltage (*Vmp*). Since *Vmp* changes mostly with temperature, it would be beneficial to have the operating voltage change accordingly. Maintaining the operating voltage near *Vmp* may require some type of voltage-regulating circuitry, as will be discussed in a later section.

Locating the System Operating Voltage

The operating voltage for a particular photovoltaic system can be determined by plotting the solar-cell output and load or storage input *I/V* curves on the same graph. This approach was first taken in Chapter 4, where we learned that the operating point of a solar cell with a resistive load is found at the intersection of the two *I/V* curves. Figure 4-12 demonstrates how a resistive load line intersects with a solar-cell curve; changes in the solar cell's curve that result from fluctuating insolation or temperature cause the operating point to trace along the resistor's linear load line. Other types of loads or storage devices can have input *I/V* characteristics that are not linear.

A basic design procedure would involve first plotting the load's input response under most operating circumstances, then sizing the PV array accordingly. The PV array's typical maximum-power voltage should conform with the load's needed operating voltage. Once the solar-cell modules have been configured to meet the load's voltage and power needs, the PV array's *I/V* curves can be plotted for different conditions of illumination and temperature. If the solar-cell and load curves are plotted on the same graph, their interaction will indicate how the system voltage will change under most conditions of insolation, temperature, and load demand.

Let us begin by plotting the input *I/V* curves for two typical types of loads, resistive and inductive, and two common types of storage, utility-interactive and battery. Figure 7-10*a* shows the *I/V* curve of a 20-Ω resistor, representative of some types of heating elements. Figure 7-10*b* depicts an inductive load (DC motor-driven air blower) *I/V* curve. In Figure 7-10*c* we see the input curve for a utility-interactive inverter with fixed-input-voltage operation. Finally, Figure 7-10*d* portrays the *I/V* input characteristics of a lead-acid storage battery. A battery's response depends on its state of charge and its temperature, and so we see a range of curves to account for these various operating conditions.

The range of voltages that will properly operate the load or storage device can be determined from their typical input curve(s); the PV array's maximum-power voltage must be within this range. A preliminary PV array design can now be made; modules are connected in series to build up the needed voltage, and connected in parallel to obtain the needed current and power. Photovoltaic *I/V* properties, however, change during different times of year as solar power and ambient temperatures fluctuate. A plot of the PV array's *I/V* curve under the extreme conditions

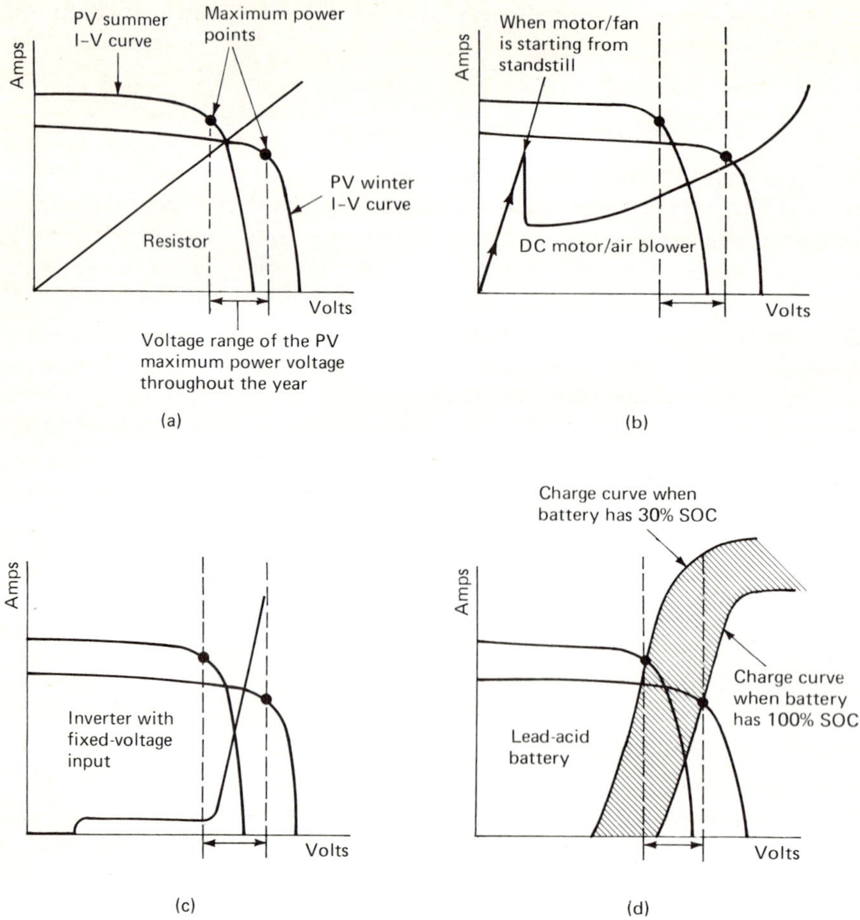

Figure 7-10 **Photovoltaic output and load or storage input current versus voltage curves: (*a*) PV output and resistor input *I-V* curves. (*b*) PV output and DC motor-driven fan input *I-V* curves. (*c*) PV output and inverter with fixed-voltage input *I-V* curves. (*d*) PV output and lead-acid battery input *I-V* curves.**

of sunlight and temperature (summer, winter, fall, and spring) will demonstrate how much the maximum-power voltage varies over an entire year (as discussed in Chapter 4); the effect of ambient temperature on cell temperature and voltage can then be predicted, and estimated *I*/*V* plots for different times of year can be made.

Sunlight varies from one site and time of day or year to the next. However, a solar cell generates an appreciable amount of power only when the insolation is high. The PV-cell temperature and maximum-power voltage in the middle of a clear day are most critical, since peak solar energy is available at that time. Cell voltage will generally be higher in the morning and afternoon because of the cooler temperatures, but will have less impact on the total photovoltaic energy output because little sunlight is available during these periods. A maximum-power voltage prediction based on the average daily cell temperature will overlook the variable intensity of sunlight, and thus will be somewhat higher than is warranted. We therefore evaluate cell temperature at a relatively high insolation level, about 80 mW/cm^2.

The maximum and mimimum cell temperatures can be calculated using the weather data for the site being considered and following the procedures and equations outlined in Chapter 4. After determining these temperature extremes, we can predict what kind of maximum-power voltage range the proposed PV array will exhibit. If we know the maximum-power voltage of the array at one temperature (which can be extrapolated from the PV module specifications provided by the manufacturer) and we also know how much the array's voltage changes with temperature, we can calculate the array's maximum-power voltage under the coldest and hottest conditions expected at the site using equations 4-2 and 4-3. If we plot the approximate PV array's *I-V* curves for these two extreme conditions on the same graph with the load or storage device input curves, we can make direct comparisons.

In each part of Figure 7-10, we see two *I-V* curves for a PV array that is suitably sized to meet the particular load or storage need. In the winter, the current output is relatively low, but the voltage output is fairly high because there is less sunlight and colder temperatures. In the summer the converse is true. Notice how the maximum-power voltage shifts throughout the year, rising in the winter and falling in the summer.

Looking at the intersection of a solar-cell array and load or storage *I-V* curve, we can determine how well the system voltage remains regulated and how well the PV array power is maximized. The shallow current-versus-voltage slope of a resistor and certain motor input curves, as shown in Figure 7-10*b*, indicates why the operating voltage of a photovoltaic system is forced to make wide fluctuations when sunlight changes. The operating voltage may be near the PV maximum-power voltage under one condition of insolation and temperature, but wander far astray under all other circumstances. In contrast, the input *I-V* characteristics of a lead-acid battery and an inverter with a fixed input voltage have a relatively steep slope. Since a PV array's maximum-power voltage varies little with changes in insolation, a load or storage device with an input *I-V* curve that also exhibits a fairly stable voltage will do a better job of maximizing PV output power (assuming that the array and load operating voltages are initially well matched). However, the battery and inverter input *I-V* curves shown in Figure 7-10*c* do not naturally follow the PV maximum-power voltage during changes in temperature. These basic PV output and load or storage input characteristics must be studied when designing a PV energy system.

Maintaining the Optimum System Operating Voltage

The optimum PV system voltage will be at or near the photovoltaic array's maximum-power voltage and within an acceptable range for operating the load or storage device. There are three fundamental ways to maximize PV output power: direct connection with proper PV-array sizing; fixed-voltage operation; and maximum-power tracking. Certain load or storage devices function adequately when connected directly to a properly sized PV array. Other devices operate better when the array voltage is fixed by using power-conditioning circuitry. The best performance is often achieved by using a maximum-power tracking circuit between the array and the load. When circuitry is added to maximize power, the system's cost increases and the overall system efficiency decreases slightly. The suitability of each of these techniques must be examined on a case-by-case basis to ascertain whether the increased power output warrants the increased cost or decreased efficiency.

Direct Connection

Certain load and storage devices have input voltage characteristics that remain relatively stable throughout most levels of operation. If a PV array is placed in a location where the ambient temperature does not vary dramatically, the PV maximum-power voltage will also remain fairly constant. In such cases, an adequate match can be made by directly connecting a PV array to its load or storage.

The lead-acid battery is a good example of a storage device that functions with a reasonably stable voltage (in fact, batteries are often used as voltage regulators). Designing a system that directly connects a PV array to a battery requires knowing the battery's charge-voltage range (battery characteristics are discussed in this chapter's battery storage section). Since the PV maximum-power voltage changes throughout the year, we will need to ascertain the output voltage range of a preliminary PV array. A hypothetical solar-cell array can be sized so that its typical maximum-power voltage is equal to the battery's average charging voltage. Normally temperature is the only factor that has a significant effect on cell voltage; therefore, the maximum and minimum solar-cell temperature at the site being considered must be estimated. If the PV array's *I-V* curve fill factor is poor (less than 0.6), changes in insolation could also have a sizable effect on photovoltaic voltage (fill factor may degrade with age, and therefore should be checked periodically).

In a battery charging system, the PV array maximum-power voltage should be equal to or greater than the battery's minimum charge voltage. We do not ever want the array voltage to be lower than that needed to charge the battery. On the other hand, we also don't want the array maximum-power voltage to be too high, because the battery would then force the array to operate well off its maximum-power voltage, and the system would not optimize the array's output capabilities. As a general rule, the PV array's lowest maximum-power voltage should equal the battery's charge voltage when it is at its lowest allowed state of charge, as shown in Figure 7-10*d* (here we are assuming that the voltage drop across the transmission line is insignificant). Using this PV and battery sizing scheme, the PV array will usually be operating at its maximum-power voltage when the battery needs energy the most, that is, when state of charge is the lowest. Conversely, when the battery is approaching full charge, the PV array current and power will naturally decrease.

Fixed-Voltage Operation

Certain power-conditioning devices have flexibility in their input circuitry design that allows for fixed-voltage operation. This feature can force the PV array to operate continuously near one voltage (which should be sized to equal the annual average maximum-power voltage). The advantage of the fixed-voltage feature is its simplicity and low cost. The drawback is that it will force the array to operate at a voltage other than the maximum-power voltage during some times of the year. If the PV array is located at a site which causes the PV maximum-power voltage to make wide migrations, a two-position "summer/winter" fixed-voltage option can be included. Even better power maximization is achieved by using a variable fixed-voltage scheme that is based on measuring the open circuit of an independent pilot cell. A PV array's maximum-power voltage remains a fairly consistent fraction below its open-circuit voltage. As the pilot cell's open-circuit voltage changes, the power-conditioning unit's control circuitry adjusts its input voltage accordingly.

Inverters are commonly equipped with fixed-voltage inputs, as shown in Figure 7-10*c*. The slope of this inverter's input *I-V* curve is not exactly vertical, but rather has a slight angle which causes the PV operating voltage to increase somewhat as the PV array power output grows. The slope angle and position of this input curve can usually be adjusted by turning one or more potentiometers in the inverter circuit.

The best fixed operating voltage for a PV array will be its average maximum-power voltage over the entire year (or half-year if two fixed-voltage settings are available). The procedure for assessing the PV array's voltage range is outlined in a previous section. The best fixed voltage is determined by checking the annual average ambient temperature at the site in question, which is found in Appendix C. In order to use Equation (4-1), we will need to know the temperature coefficient of our particular PV array. Since this constant depends on how the PV modules are mounted, for the most accurate determination a sample module should be tested in the proposed mounting configuration. The typical cell temperature (which is weighted toward periods of high insolation) at the average daytime ambient temperature and an insolation of 80 mW/cm^2 is then calculated using Equation 4-1. Knowing the typical cell temperature over an entire year enables us to ascertain the average maximum-power voltage; the fixed-voltage setting of an inverter's input must be designed to roughly equal the above-determined array voltage. This setting can and should be fine-tuned to achieve actual measured maximum power when the inverter is finally installed in the system.

Maximum-Power Tracking

Certain loads have operating voltages that can be quite erratic and would do a poor job of tracking a PV array's maximum-power voltage. A photovoltaic array can sometimes be placed in a location that will expose it to very wide temperature fluctuations; as a result, the PV maximum-power voltage will vary greatly, and significant power losses will be suffered when the direct-connection or fixed-voltage techniques are used.

A DC motor driving an air blower has an input *I-V* characteristic that is generally incompatible with a normal PV output performance. The motor's rotation speed and the blower's air displacement rate are proportional to input voltage; consequently, the motor's voltage fluctuates as far as is necessary to maintain the highest speed possible. A motor's starting power requirements also pose some problems. Let us examine the input *I-V* curve of the DC air blower as shown in Figure 7-10*b*.

A DC motor requires a significantly higher current to start turning than it does to continue turning. This fact is demonstrated by the narrow crest in the curve in Figure 7-10*b*. During the morning or evening hours of a sunny day or throughout some cloudy days, a PV array may often have enough power to run the motor, but will not have enough power to start it from a standstill; the motor does not move, and the array power is not put to use. This situation is difficult to remedy. When a motor is not moving, it does not build up a countervoltage, and it therefore short-circuits whatever power source is attached to its input. A PV array that is forced to operate near its short-circuit current is delivering far less than its maximum-power capability. It is also possible for a PV array to have enough power to start a motor, but still be forced to operate far below its maxium-power voltage and thus to be hamstrung by the motor's input characteristics. This problem can be solved by oversizing the current-output capability of the PV array or by using a maximum-power tracker.

A maximum-power tracker is placed between a PV array and its load, and is based on a design that samples the PV output and changes the apparent impedance of the load until the PV power is maximized. PV output power can be sampled either by multiplying the PV voltage and current readings together or in a special way by measuring just the voltage or current. The apparent load impedance as seen by the array is changed, and the PV power output is sampled continuously to determine whether the power is increasing or decreasing. Whenever power is diminishing, the impedance sizing circuit is instructed to change direction (for example, if the impedance size is decreasing, it will be switched to increasing). On the other hand, if the power is being augmented, the impedance sizing circuit is permitted to continue in its present trend. A maximum-power tracker is always in flux as it hunts for maximum power, because it has to overshoot its mark to confirm it.

A maximum-power tracker can be treated as a black box; only basic specifications need be known. Its input/output efficiency should be no less than 90% to 95%. The device has a voltage operating range and maximum power and current capability that must be sized for each application. Up to a point, the wider the range of voltages that can be tracked, the higher the cost and the greater the maximum-power tracking effectiveness. A trade-off develops that must be tested and resolved. A maximum-power tracker can be designed to obtain its operating power either from an external source or from the direct current provided from the photovoltaic array.

The need for maximum-power tracking is increased when a load or power-conditioning device has an unstable input voltage or when the PV array is operating in a location where large temperature extremes are experienced. When possible, fixed-voltage operation that suitably accounts for local temperature and insolation conditions has generally proven to be most economical and practical. Under direct comparison, the extra energy obtained by maximum-power tracking is usually not significant enough to warrant the small efficiency losses and increased initial and maintenance costs. The best fixed-voltage setting, however, must be accurate to be effective.

DIRECT-CURRENT-TO-ALTERNATING-CURRENT CONVERSION

Electricity is generated and used in two distinct forms: direct current and alternating current. The relative advantages of forcing electrons to uniformly flow in one direction versus cycling them back and forth in alternating directions have led to varied applications throughout recent history. The diverse needs of an industrialized economy sometimes favor DC and other times AC, depending on their respective features.

The first electric power plants generated DC electricity and were designed by Thomas Edison and St. George Lane Fox in 1878. Large power losses were experienced, however, when they attempted to transmit direct current over long distances; consequently, other forms of electricity were explored. Alternating-current generators, invented by two Englishmen in 1881, provided a solution to the transmission problem. By stepping voltages up very high during transmission, power losses in the lines could be minimized. Alternating current could easily be stepped up to high voltages by a simple and efficient device called the *transformer,* while no technology was available to change DC voltage levels. Ever since the end of the nineteenth century, the application of AC electricity has predominated throughout the world. Only special industries, automobiles, airplanes, and boats have traditionally relied on DC, because DC can easily be stored in batteries or used to perform specialized functions such as electrolysis or DC motor operation.

Recent technological developments involving diodes have made it possible to alter DC voltage levels. This discovery and certain advantages of DC transmission have led to a renewed interest in DC electricity. Photovoltaic cells naturally generate direct current; if solar cells become cost-effective and achieve widespread use, industry will undoubtedly be spurred to market more machines and appliances that function with DC. Until that time, the momentum of industry and certain AC features will favor the continued use of alternating current, and thus DC-to-AC conversion of photovoltaic electricity will often be required.

The Relative Advantages of DC and AC Electricity

The relative advantages of AC and DC electricity are worth examining. There are simple techniques for transforming AC voltage levels, while DC voltage adjustments are somewhat more difficult to perform. On the other hand, alternating current has reactance associated with it that can lead to stray losses of energy during transmission, while direct current has only wire resistance to contend with. Certain machines and devices can function only on either DC or AC, and therefore are dedicated to a particular form of electricity.

The major advantage of AC electricity is that important voltage-level changes can be made simply, cheaply, and efficiently. When transmitting electricity over long distances, it is necessary to step the voltage up very high, since line power losses and wire costs are less at high voltages than at high current levels ($P = V^2/R$ versus $P = I^2R$). When the electricity reaches its destination, it must be stepped down to a practical level. Most loads operate at voltages between 5 and 440 V, while during transmission voltages are raised as high as 350,000 V. AC voltages are changed by a transformer, which harnesses the magnetic field produced by alternating current to induce an appropriate voltage in another conductor. In contrast, DC electricity produces no changing magnetic field and thus is incapable of inducing voltages at any level. DC voltage can only be altered either by converting DC to AC, performing a voltage transformation, and then rectifying back to DC, or by using an electronic circuit called a DC-to-DC converter (which also involves some type of oscillator). AC circuits therefore afford the simplest means of producing important changes in voltage.

The major disadvantage of AC electricity is its tendency to interact with conducting materials in the vicinity of its path in such a way that reactive power is generated and partially lost during transmission. Reactance can be either inductive or capacitive. Inductance is generally created when the changing magnetic field around an AC conductor induces an opposing voltage in another part of the original wire. When a conductor is looped in a coil, inductance is heightened because the magnetic field is positioned in such a way that it readily induces a voltage in the circuit. The nature of inductive reactance is to oppose sudden changes in current. Capacitance is created when any two adjacent conducting materials are separated by an insulating material. When a difference in voltage potential exists between adjacent conductors or metal plates, energy is stored in the electric field built up in the insulating separator. As the metal surface area is increased or the thickness of the insulating gap is decreased, the area and strength of the electric field are augmented. This field produces capacitive reactance, which by nature opposes sudden changes in voltage. There are both desirable and undesirable reactances in every circuit, the latter being unavoidable and a source of problems.

When AC electricity is flowing, the current sine wave lags in time behind the voltage sine wave if the circuit is inductive, while the voltage sine wave lags behind the current sine wave in a capacitive circuit (the amount of current or voltage lag depends on the size of the inductance, capacitance, resistance, and the AC frequency). When one sine wave is out of synchronization with another, their respective peaks and zero crossings do not coincide, and they are said to be *out of phase,* as shown in Figure 7-11. Inductance and capacitance have opposing effects in a circuit, and hence can cancel each other out if they are equal in magnitude. For this reason, the utility often puts capacitors on part of its transmission line to reduce the largely inductive reactance that exists in most transmission and load circuits. Figure 7-11 depicts the size and timing of current and voltage waves in different reactive circuits.

The net effect of phase shifting between AC voltage and current is that the *real power* used by the load is only a portion of the power generated by the power supply. Where does the remaining power go? Inductive and capacitive reactance do not dissipate energy in themselves; they simply cause energy to temporarily be stored in a magnetic or electric field. This temporary storage of energy forces the AC generator to produce more than the load requires and causes unused energy to be idly transferred back and forth between the generator and the load. Part of this *reactive power* moving back and forth in the line is dissipated in the wire resistance, thus increasing power-line losses or costs.

The power factor rating of a circuit is the ratio of real power being dissipated at the load to the total power being produced by the power supply (a power factor of 1.0 indicates that there is no reactive power, while a PF of 0.5 indicates that half of the total power produced is reactive). Since reactive power serves no useful function, circuits should be designed with as high a power factor as possible. Figure 7-11 shows the waveforms of circuits with different power factors. Notice how the current amplitude and total power increase as the reactive power increases and the power factor diminishes.

By its nature, there can be no inductive or capacitive reactance in a DC circuit, and therefore power output and line power losses are kept to a minimum. Wires and cables in a DC circuit need not be oversized to handle additional reactive power.

Certain machines and appliances actually require DC electricity to operate, while others make effective use of AC. Direct current is necessary for various functions. For instance, electronic circuits, found in many types of apparatus, including computers, televisions, stereos, and radios, are able to operate only on DC (the AC line voltage usually used to operate these devices has to be stepped down and converted to DC using a rectifier circuit). Industrial functions involving electrolysis or

---

Figure 7-11 **Power factor and the phase relationship of alternating current and voltage: (*a*) an ideal case where voltage and current are exactly in phase; (*b*) a case where voltage and current are slightly out of phase; (*c*) a case where voltage and current are moderately out of phase; and (*d*) a case where voltage and current are extremely out of phase. (Charles Cox, "Making the Utility Connection," *Solar Age*, Harrisville, N.H., December 1981, p. 43.)**

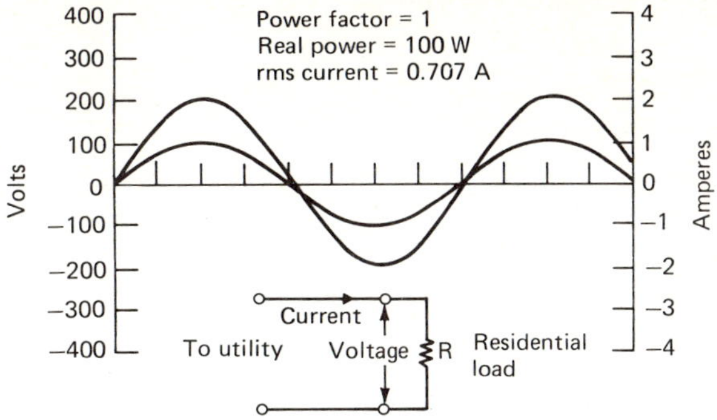
Power factor = 1
Real power = 100 W
rms current = 0.707 A
Volts
Amperes
Current
To utility
Voltage
R
Residential load

(a)

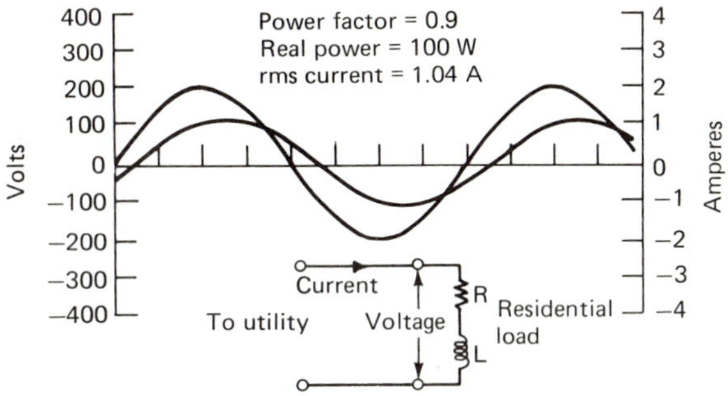
Power factor = 0.9
Real power = 100 W
rms current = 1.04 A
Volts
Amperes
Current
To utility
Voltage
R
L
Residential load

(b)

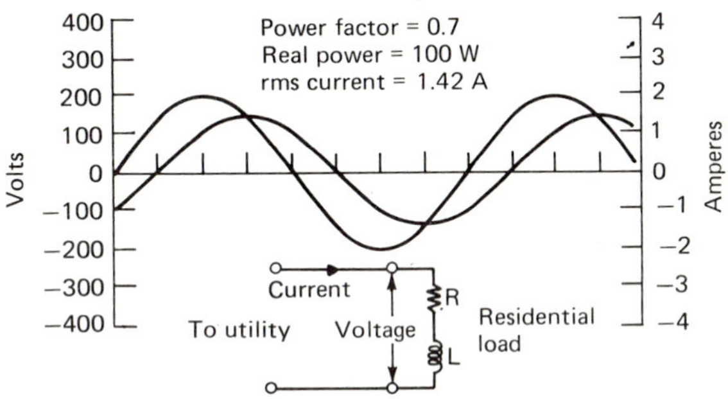
Power factor = 0.7
Real power = 100 W
rms current = 1.42 A
Volts
Amperes
Current
To utility
Voltage
R
L
Residential load

(c)

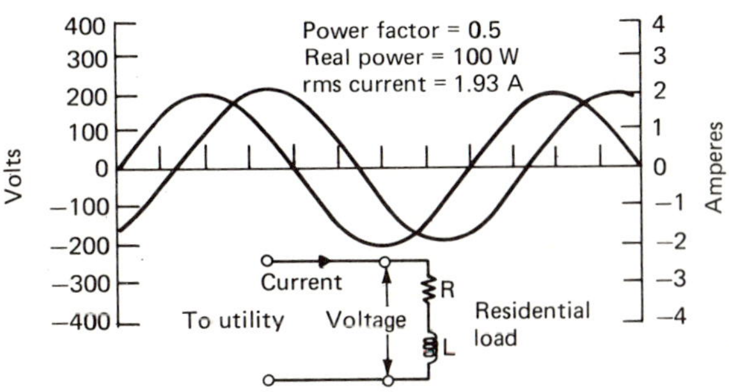
Power factor = 0.5
Real power = 100 W
rms current = 1.93 A
Volts
Amperes
Current
To utility
Voltage
R
L
Residential load

(d)

electroplating also require DC. DC motors are used in some industrial applications and in electric-powered vehicles because of their intrinsic high starting torques (enabling quick acceleration). AC motors are generally smaller for a given amount of output power, less expensive to manufacture, and require less maintenance than DC motors (DC motors have brushes that need to be replaced periodically). Three-phase AC motors are used in industrial applications because they are more efficient and less expensive, even though bringing in three-phase utility power is itself an expense (three-phase AC, provided by the utility for major customers, is in the form of three separate AC signals that have different timing and thus are said to be proportionally out of phase). Some appliances ingeniously utilize the AC waveform. An AC electric clock, for instance, gets its timing from the precise 60-cycle-per-second waveform; silicon-controlled rectifier light dimming and motor speed controls use the changing AC voltage to perform variable switching.

A variety of issues must be considered before deciding which form of electricity to use. The advantage of DC electricity is that it has no reactive power losses during transmission; the advantage of AC electricity is that AC voltages can be changed more easily to meet the voltage requirements for transmission and various load applications. The conversion from DC to AC is more difficult and costly than the opposite. Rectifying AC into DC usually requires a few inexpensive diodes and a capacitor. Techniques for generating AC from DC involve more parts and costs, as will be discussed in the following section. Consequently, DC becomes more cumbersome if voltage levels must be changed often. Within a particular residential, commercial, or industrial establishment, appliances may be operating at various voltages. High-power loads, such as those involving large heating or cooling functions, usually operate at a high voltage to keep the current and power losses down during transmission. Low-power loads, on the other hand, are more efficient at or require low voltages. Alternating current allows these multiple voltage transitions. Independent power systems, such as those found in automobiles, airplanes, and boats, are almost always driven with direct current, since they involve low-power loads that can all operate at one uniform voltage and they usually use batteries for storage and voltage regulation.

A photovoltaic array naturally generates DC electricity. A photovoltaic system, therefore, is simplest when the load can be powered with direct current. If all machines and appliances can operate off of one voltage, an all-DC power system is an attractive option. The designer can forgo the cost and complexity of a DC-to-AC inverter. Appliances exist or can be developed to perform most functions in a residential or commercial establishment with DC electricity. DC-powered apparatus, however, are generally more expensive than their AC counterparts because at present they are not mass-produced in the same large numbers. The higher cost of DC appliances is usually weighed against the cost of a DC-to-AC inverter before deciding which course to take.

The Basic DC-to-AC Inverter Circuit Design

The conversion of the direct-current electricity produced by solar cells into the alternating current used by most appliances and generated by major utilities poses three basic problems. First, the efficiency of the DC-to-AC conversion process must be as high as possible; excessive losses of energy will make the process uneconomical. An inverter efficiency of less than 85% generally is considered prohibitive. Second, a utility-interactive inverter has an inductive reactance and a lagging power factor that act as a load on the utility. This reactance must be minimized, and the power factor should be between 0.8 and 1.0. Third, some inverters produce harmonic distortion or harmonic waves greater than 60 cycles per second that might damage reactive loads connected to the circuit. Harmonic distortion is the ratio of undesirable harmonic power to normal 60-cycle-per-second power. Little is presently known about the effects of harmonic distortion on the utility grid; the maximum tolerable harmonic distortion for a utility-interactive inverter is estimated to be about 10%. The three basic issues that must be addressed when designing or choosing a DC-to-AC inverter, therefore, are efficiency, power factor, and harmonic distortion.

Sinusoidal alternating current is naturally produced by the circular motion of a conductor through a magnetic field, as seen in an electric generator with its rotating armature. Rotation of the armature is generally achieved by forcing steam or water across the blades of a propeller connected to its shaft. The output of an AC generator is a pure sine wave that is centered on the zero amplitude axis.

The sine wave is a building block of periodic motion. As revealed by Fourier analysis, almost any periodic wave that is not a sine wave can be broken down into constituent sine waves, called *harmonics,* which have a period (time for one cycle) that is some fraction of the major or fundamental wave. In other words, any nonsinusoidal repetitive wave usually is the sum of various sine waves of higher frequency, varying amplitude, and varying phase relative to the fundamental. This fact is demonstrated in Figure 7-12, where a triangle wave is produced by adding harmonics of the fundamental whose frequency and amplitude have been calculated using Fourier analysis (the fundamental is the major wave which has the largest amplitude and lowest frequency; in this case the fundamental has the frequency and amplitude of the triangle wave).

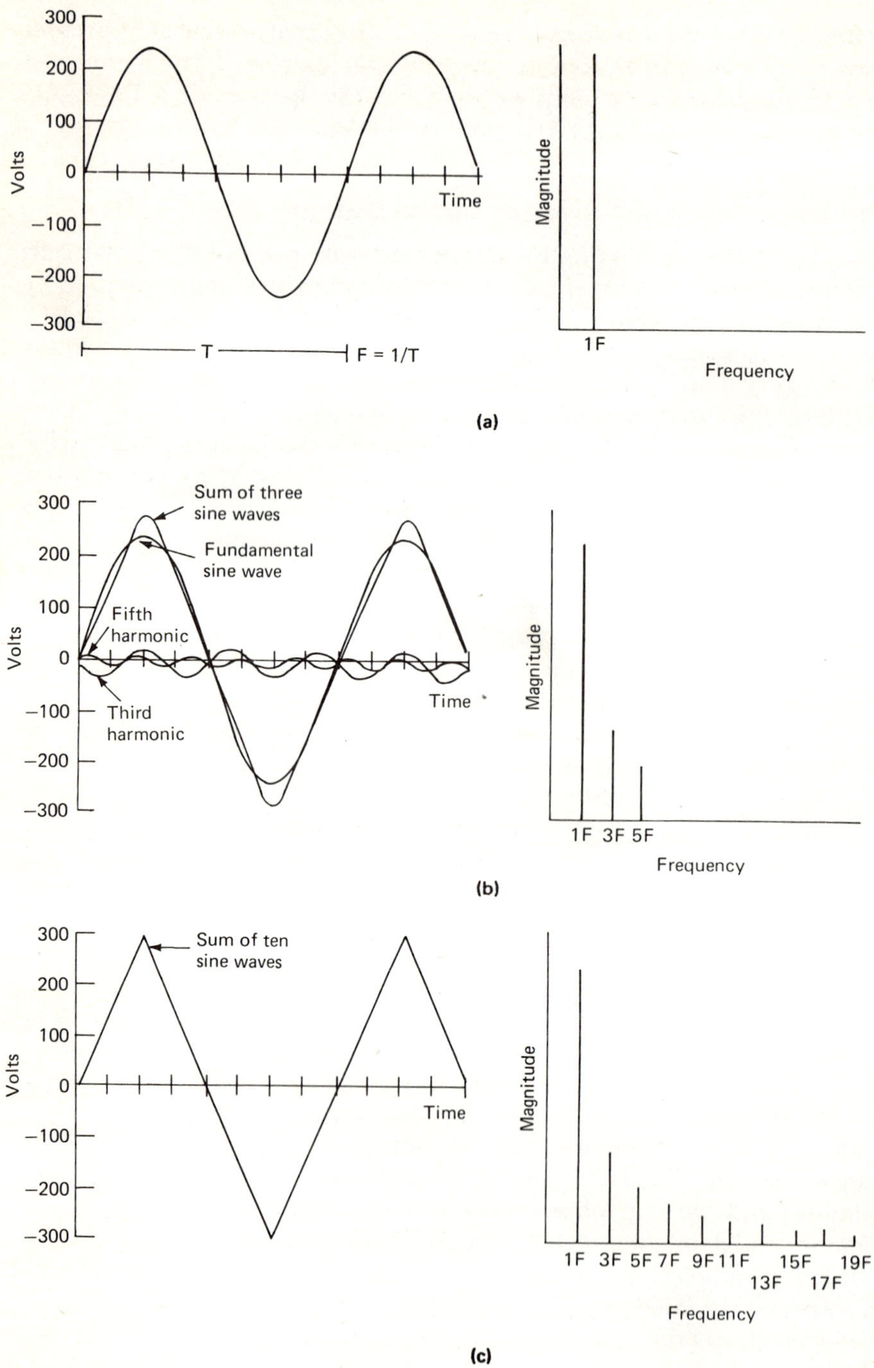
300
200
100
0
−100
−200
−300
Volts
Time
T
F = 1/T
Magnitude
1F
Frequency
(a)
Sum of three sine waves
Fundamental sine wave
Fifth harmonic
Third harmonic
Time
Magnitude
1F 3F 5F
Frequency
(b)
Sum of ten sine waves
Time
Magnitude
1F 3F 5F 7F 9F 11F 13F 15F 17F 19F
Frequency
(c)

A simple way to generate a pure 60-hertz (cycles per second) sine wave from a DC source is to use a motor generator. A photovoltaic array can power a DC motor, which in turn drives an AC generator. The output waveform is sinusoidal with no harmonic distortion and a high power factor. The major drawback of such a design is its low conversion efficiency. The efficiencies of the motor and alternator are multiplied together to produce a total efficiency of between 50% and 80%, which is well below the suggested limit of 85%.

Electronic circuitry has been developed that performs DC-to-AC inversion with high efficiency; the sinusoidal waveform, however, has to be synthesized using sophisticated electronic logic. The process of converting DC to AC with a static inverter (involving no moving parts) involves three basic stages: electric polarity cycling, voltage regulation, and voltage or current sine-wave shaping. The DC polarity must be alternated back and forth at 60 cycles per second; the output voltage must be regulated at a fixed voltage, which usually will be 120 or 240 V AC; and the voltage or current waveform must be controlled so that it most nearly approximates a sine wave.

The full bridge inverter circuit, shown in Figure 7-13, is the basis for most DC-to-AC inverters. As pairs of switches (pair S_1 and S_2 or pair S_3 and S_4) are alternately opened and closed, current through the load flows in one direction and then the other; hence we have alternating current. The switching device must be able to open and close rapidly and must dissipate very little power in order to maintain a high inversion efficiency. The semiconductor revolution has developed various switching devices, such as the transistor or the silicon-controlled rectifier (SCR), that effectively perform this function.

In the United States, utility current alternates 120 times a second (60 Hz with two alternations per cycle). In order to copy the utility AC pattern, the pairs of bridge switches will similarly have to alternate 120 times a second. There are two basic ways of controlling the switching operation: line commutation and self-commutation. The switches in a line-commutating inverter are triggered by the 60-cycle timing of the utility, while the self-commutating inverter switches are driven by a clock-

---

Figure 7-12 Any periodic motion can be broken down into constituent sine waves. (*a*)Two equivalent ways to represent a sine wave are the time plot *(left)* and the spectrum plot *(right)*. (*b*) Adding the right amounts of third and fifth harmonics to a fundamental sine wave generates the waveform labeled *sum of three sine waves*. (*c*) After ten harmonics have been added to the fundamental sine wave, agreement with the original triangular-shaped waveform is reasonably good. All the components have been included in the spectrum plot at right; only the final result is shown on the time plot. (Charles Cox, "Making the Utility Connection," *Solar Age*, December 1981, p. 44.)

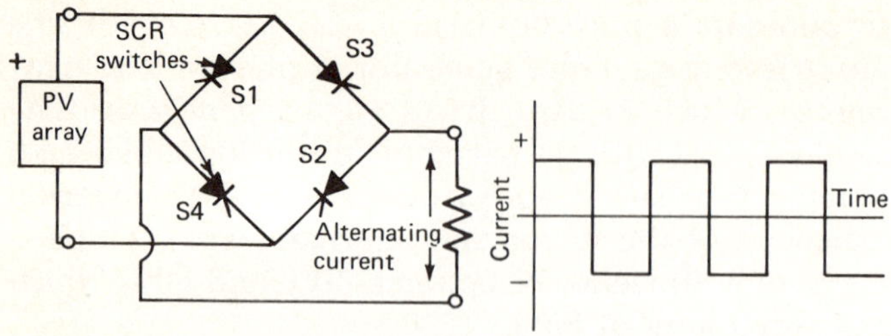

Figure 7-13 **The full bridge inverter circuit with its square-wave output.**

ing source internal to the inverter. A line-commutating, utility-interactive inverter is less expensive but is dependent on the utility to operate. A self-commutating inverter's price is increased by the additional circuitry needed to perform self-commutation, but it has the capacity to operate in a stand-alone mode.

The output wave shape of a basic inverter is not a sine wave as desired, but rather a square wave, as shown in Figure 7-13. The efficiency of the full bridge inverter is high (between 90% and 100%), but the harmonic and wave-shape distortion is abysmal. Engineering involves trade-offs that have to be balanced; solving one problem often leads to another. The 60-Hz square wave is the aggregate of a 60-Hz sine wave and a wide range of higher-frequency harmonic sine waves. Certain AC loads can function adequately on a square wave. Apparatus involving heating elements or incandescent lighting do not care whether the alternating current is sinusoidal or square-shaped. Other AC loads can be damaged if they are powered with a square wave. In particular, instruments involving transformers and motors will tend to overheat because they absorb some of the wide range of harmonic energy that exists in a square wave. To make the inverter output appropriate for all types of loads, it will be necessary to filter out undesirable harmonics.

Filtering and shaping the inverter's square-wave output into the requisite sine wave is accomplished in a variety of ways that involve discrete components and high-speed switching circuits. Many techniques used to achieve wave shaping involve placing a transistor switch between the photovoltaic power source and the bridge circuit. The transistor is switched on and off at a high frequency (between 1 and 20 kHz) and for varying durations of time that depend on comparison with or control from a reference sine wave. This reference wave can be generated either from internal logic and memory circuitry or from an external utility AC signal. The inverter's AC output wave form is further improved by filtering with discrete components such as inductors and capacitors (inductance in the inverter's output will have the detrimental effect of decreasing its power factor, and therefore must be minimized or counterbalanced).

The inverter's root-mean-square (rms) voltage output over many AC cycles has to be stable, even though the PV array's voltage and current will fluctuate with sunlight intensity and ambient temperature. An inverter's output is typically regulated at 120 or 240 V rms, while its input from the PV array may fluctuate dramatically. A stand-alone inverter controls its output voltage while allowing its power and current to follow the load demand. A utility-interactive inverter has its voltage anchored to the utility's voltage and controls its current and power output according to the available solar power. Voltage or current regulation is generally achieved by using a control circuit wih a feedback loop that senses the inverter's input and output and adjusts the voltage or current drawn from the PV array to meet the output requirements.

A commercially available, off-the-shelf inverter can be treated as a black box; only basic specifications need be known. The inverter will have either stand-alone or utility-interactive capabilities or both. The maximum power rating, the acceptable DC input voltage range, and the rated AC output voltage are important parameters of the inverter that must be known in order to integrate it into a photovoltaic energy system (an inverter's maximum power should be rated in watts, a volt-ampere rating will read deceptively high if the inverter's power factor is low, since the voltage and current sine waves are out of phase). An inverter will employ one of various techniques to maximize the power output of the PV array; the advantages and disadvantages of these different methods must be considered (this issue is discussed in the voltage regulation section). What kind of power factor and harmonic distortion does the inverter have under normal operating conditions? Is the inverter's input electrically isolated from the output with a transformer so that a combined ground in the PV array and the utility circuit will not short out half of the inverter's output cycles? Finally, what is the inverter's DC-to-AC conversion efficiency?

The efficiency of a DC-to-AC inverter is not continuous throughout all power levels of operation. The efficiency/input-power curve for a common utility-interactive inverter, as shown in Figure 7-14, indicates that efficiency drops off as the output power decreases from the inverter's rated output. The *total harmonic distortion* (THD) depends on the power level and on how well the sine-wave simulation or filtering of harmonics is performed; it typically is held below 5 to 10%. An inverter's power factor is determined by the amount of inductance in the inverter's output circuitry and also can vary with the level of power input; the power-factor rating should be between 0.8 and 1.0 (at rated power output). If the input and output of an inverter are not isolated electrically

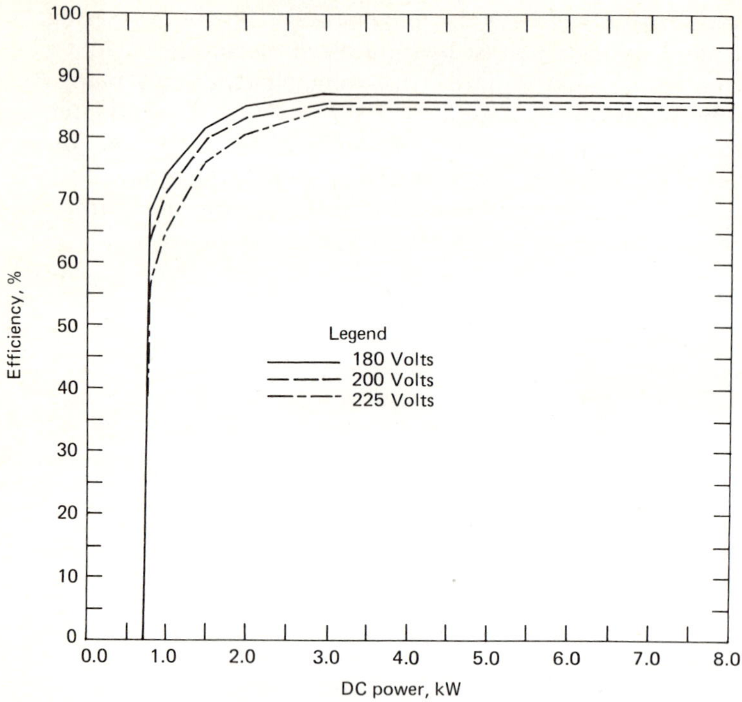

Figure 7-14 Efficiency/input-power curve for a self-commutating inverter in the utility-interactive mode; a 6-kVA Abacus at NERES. (M.I.T. Lincoln Laboratory.)

with a transformer, it will not be possible to attach the array circuit to ground; less expensive but also less reliable methods will have to be used to prevent accidental injection of DC power into the utility grid or AC power into the PV circuit. Eliminating the transformer, however, will increase the inverter's efficiency and reduce its cost.

An inverter is generally sized to have a peak output rating that roughly equals or is a little greater than the typical peak power of the PV array (which varies from one site to the next); more precise sizing requires knowing the average peak power output of the array and the precise efficiency/input-power curve of the inverter. The DC input-voltage range of an inverter must be observed; voltages outside this range will cause the inverter to malfunction or stop operating. The PV array's maximum open-circuit voltage on the coldest day and the minimum array operating voltage must be within this range. Clearly, the inverter's AC output voltage must be compatible with the utility or loads that will be serviced. Various techniques for maximizing the power output of the array can be incorporated in the inverter's input circuit: maximum-power tracking, fixed-voltage operation, and fraction-of-open-circuit-voltage tracking (these approaches are dealt with in the voltage regulation section).

The aforementioned characteristics of an inverter should be listed in the manufacturer's published specifications. Designing a utility-interactive PV system will involve choosing an inverter with features that make it compatible with both the utility and the PV array.

8

SYSTEM DESIGN PROCEDURES

In this chapter we will develop a step-by-step approach to designing a photovoltaic system. An overall equation is needed to calculate the PV-system energy output which accounts for the losses that naturally occur in the energy conversion and power-conditioning process. The magnitude and timing of the PV energy output and the load energy input must be coordinated. Sizing a solar-cell system to satisfy a particular energy demand will be facilitated if a basic procedure and method is available. The cost-versus-benefit appraisal of a solar electric generator will be crucial; a simple and reliable economic evaluation program will assist people in making wise investment decisions.

PHOTOVOLTAIC-SYSTEM ENERGY CALCULATIONS

There are energy losses between the output of a photovoltaic array and the output of a photovoltaic system that must be accounted for when calculating the total solar electricity that can be generated. In Chapter 5 we developed Equation (5-1) to calculate the photovoltaic-array output given solar-energy data for the locality under consideration. Since then we have seen the need to add various types of power-conditioning or storage equipment which, in the process of operation, result in small losses of energy. It is also apparent that energy is lost in the photovoltaic-array wiring. The rated efficiency of a solar-cell array is measured when it is operating at its maximum-power voltage. In many cases, the system will operate for part of the day or year at a voltage other than the optimum, which will also lead to power losses. Equation (8-1) accounts for the major predictable losses between the array and its load.

$$PSE = PE \times PCE \tag{8-1}$$

where PSE = photovoltaic system energy for one entire day (kWh/day)
PE = photovoltaic energy for one day as calculated using Equation (5-1)
PCE = power-conditioning efficiency, averaged over one day, accounts for losses in power-conditioning equipment such as an inverter, a battery, or a maximum-power tracking or control circuit. If the system has more than one power-conditioning device, multiply efficiencies together to obtain total PCE (for example, 90% × 85% = 76.5% = 0.765)

The efficiency of a power-conditioning device can be either measured or obtained from the device's specifications. Attention must be paid to the conditions under which the efficiency is measured.

An inverter's efficiency varies with input power (as shown in Figure 7-14), and therefore the daily efficiency depends on the inverter's average input power throughout a typical day. An inverter and array are usually sized to operate at an average power level of about 60% to 100% of the inverter's rated power; the efficiency at that power level typically is between 80% and 95%.

Battery efficiency is best represented by the battery's rated in/out watthour efficiency. As discussed in a previous section, battery efficiency varies with operating temperature, discharge rate, and storage time. If the battery is expected to operate in less than 0°C temperatures for extended periods, or if the discharge rate is not properly matched with the battery's optimum rate, the efficiency will be less than specified. Typically, a lead-acid battery's watthour efficiency ranges between 80% and 95%.

The efficiency of a maximum-power tracking or control circuit is generally very high, in the neighborhood of 95%. This efficiency will vary somewhat with input power, since the ratio of power required to operate the circuit divided by the array's output power increases as sunlight decreases, and vice versa.

There are two other factors that will result in power losses between the photovoltaic array and its load that are not included in Equation (8-1). First, a photovoltaic system operating without a maximum-power-point tracker will sometimes not deliver its rated and optimum power because the system operating voltage will not always match the array's maximum-power voltage. Second, wire resistances and voltage drops across array-circuit diodes also result in power losses. The latter two considerations will either be too difficult to predict or should be negligible (less than 3%), and therefore will not be taken into account.

An accurate prediction of how much power is lost as a result of poorly matched operating and maximum-power voltages involves establishing detailed computer models of the proposed system that require extensive weather data with hourly readings (like those found for 26 sites on the SOLMET tapes). Accurate predictions of the hourly load energy demand will also be needed. The system designer may not have access to the facilities, money, and time needed to make such a determination. Detailed weather data, averaged over many years, are also not available for many parts of the country. The unmaximized power losses will normally be minimal for a utility-interactive system with fixed voltage or maximum-power-tracking operation, and will be somewhat larger for a battery storage system with its wider range of operating voltages (not to mention the energy losses due to battery-protecting array shedding or power dumping). If the PV array, power-conditioning equipment, and load are suitably matched in size, we can assume that the power-maximizing losses will be reasonably small.

The major wiring power losses will take place in the blocking diodes and the wire resistance. If a system has blocking diodes in each string or at the array output, the fraction of total power lost will be about 0.6 V divided by the array's maximum-power voltage ($0.6/V_{mp}$). Clearly, the higher the array operating voltage, the less significant the diode power losses. The array-circuit wire resistance depends on the type of wire used for the application. Chapter 9 discusses the procedure for choosing the proper wire gauge and has wire resistance tables to enable the designer to make power-loss calculations for different types of wire. In any case, if a system has a relatively high operating voltage (greater than 60 V), and has properly sized the circuit wiring, the wiring losses should be negligible and therefore need not be considered.

PHOTOVOLTAIC-SYSTEM SIZING

We shall follow a step-by-step approach to designing a photovoltaic system. The designer is presented here with a procedure for sizing the various components that will play an integral part in satisfying the demand for energy.

1. Determine the typical load electric energy demand on an average day for each month of the year (kWh/day). If monthly data are not available, then choose four different times of the year, preferably near the solstices and equinoxes (December 21st, March 21st, June 21st, and September 21st). Plot the energy demand (kWh/day) versus time of year (month).

2. Make a preliminary decision as to whether the photovoltaic array will be sized to satisfy the annual peak, average, or minimum energy demand shown in the graph drawn in step 1. If needed, calculate the annual average daily energy demand (kWh/day).

3. Determine the total solar energy (TE) available at the site under consideration on an appropriately tilted south-facing surface (as discussed in Chapter 3). Adjust the solar energy received on a horizontal surface to that on a tilted surface using the equations or figures in Chapter 3 or the tables found in Appendix C. Establish a table of daily solar energy versus month or season of year (kWh/m^2 · day versus time of year).

4. Make a preliminary decision as to whether the array is going to be sized according to the annual peak, average, or minimum solar energy. If needed, calculate the annual average daily solar energy per tilted square meter from the table compiled in step 3.

5. Complete a preliminary system design by choosing the type of power-conditioning equipment needed. Will the system be utility-interactive? Will it have battery storage? Will it involve a maximum-power tracking or control circuit?

6. Determine the approximate photovoltaic array size needed to meet the energy demand estimated in steps 1 and 2. Based on the decisions made in steps 2 and 4, match up the annual peak, average, or minimum daily energy demand with the annual peak, average, or minimum daily solar energy received. The array area is calculated using Equation (8-2).

$$\mathrm{AA} = \frac{\mathrm{LE}}{\mathrm{SE} \times \mathrm{ME} \times \mathrm{TC} \times \mathrm{PF} \times \mathrm{PCE}} \tag{8-2}$$

where AA = array area (m^2)
LE = one day's load energy demand (kWh/day) based on the annual peak, average, or minimum
SE = solar energy for one day at site under consideration (kWh/m^2 · day), based on annual peak, average, or minimum
ME = module efficiency at NOCT or local cell temperature

TC = efficiency temperature correction factor to adjust module efficiency to NOCT or local cell temperature, if not already there (see Figure 4-4 if necessary)
PF = packing factor for module or array (active cell or module area/total module or array area). *Note:* ME will already account for cell packing factor; if so, only use module packing factor.
PCE = power-conditioning efficiency for power-conditioning equipment such as inverter, battery, or maximum-power tracker

Dividing the daily energy demand (kWh) by the daily photovoltaic energy output (kWh/m^2) will yield the array area in square meters.

7. Calculate the peak power rating of the array sized in step 6 by multiplying the array area (in m^2) by 1000 W/m^2 and by the cell efficiency and array packing factor as performed in Equation (5-3).

$$PP = A \times 1000 \text{ W/m}^2 \times ME \times PF \tag{5-3}$$

where PP = array peak power at NOCT
ME = module efficiency at NOCT

8. Choose particular photovoltaic modules available on the market and configure them into an array composed of strings and branches that will meet the needed peak power and output voltage requirements (the module's rated power and voltage must be known). Based on the temperature and insolation extremes experienced at the site, determine the maximum-power voltage range of the array (using Equation 4-2) and determine whether this scale is compatible with the load or power-conditioning device. Arranging the modules in series and parallel (as discussed in Chapter 5) to produce a desired voltage at the needed power level can be difficult if the module's voltage and power output are not a convenient value; in such a case, a different module can be chosen or the array may have to be over- or undersized.

9. Calculate the daily energy output of the proposed photovoltaic system by using the average daily insolation data for a tilted surface, developed in step 3, and the solar-cell array and system Equations (5-1) or (5-2) and (8-1). Plot the PV-system energy output versus time of day on the same graph drawn in step 1.

10. Examine the PV output and load input curves in the graph completed in step 9 and note the amount and times of photovoltaic-energy surplus and deficit throughout the year. If the surplus or deficit is

excessive, change one or more of the design parameters and repeat some or all of steps 1 through 9. Will long-term storage or backup power be needed?

11. Estimate the load power requirements (W) throughout a typical spring or fall day or an annual average day. Draw an approximate plot of load power (W) versus time of day for a 24-hour period (hr).
12. Estimate the solar power available at the site being considered during different hours of a typical spring or fall day or an annual average day (W/m^2). Multiply the array peak power (as determined in step 7) by the annual average percentage of possible sunshine (as found in Table 3-2) to obtain an approximation of the annual average midday power level. Consulting an almanac or using the nomogram in Figure 3-15, determine the hours of sunrise and sunset for a spring or fall day or a declination of zero. At sunrise and sunset the array power level is just reaching or leaving zero. Plot the photovoltaic power output versus time of day on the same graph established in step 11 by using the three data points determined above and drawing a half sine-wave curve as depicted in Figure 3-12.
13. Examine the PV output and load input curves in the graph completed in step 10 and note the amount and times of photovoltaic-energy surplus and deficit throughout the day. Will short-term storage or backup power be needed?

Battery Storage

In a stand-alone system with battery storage, the photovoltaic array is usually sized so that its energy output throughout the year is never less than the load's energy demand (that is, the annual minimum PV-system energy output equals the annual maximum load energy input). The system also must maintain a continuous energy supply at night and on cloudy days although there is little or no solar energy. The amount of battery storage needed will depend on the load energy demand and on weather patterns at the site. The percentage of possible sunshine for the site, found in Table 3-2, will give a general idea of what fraction of any month or year has overcast skies. If the annual percentage is 50%, for example, we know that half of the year experiences cloudy weather. We need to know, however, how the cloudy weather is distributed. The sunshine percentage for each month and any available data indicating the longest continuous period without direct sunshine will help the designer predict how much storage is needed.

Having sufficient storage and photovoltaic output for the most overcast and lowest insolation time of year will result in excess storage capacity and photovoltaic energy for the least overcast and higher insolation periods. During peak solar energy days, the battery will remain near a full state of charge and will rarely need to dip deeply into its reservoir. At the same time, the array will be putting out a surplus of energy that will have to be disconnected or diverted from the battery to prevent overcharging. For example, notice the curves in Figure 8-1, which are based on the operation of the WBNO radio station photovoltaic system with battery storage described in the Appendix. In this figure, we see that the array output current drops off and array power is discarded when the battery voltage reaches its maximum allowed level. At this point the battery protection circuit has been forced to shed a large portion of the PV array.

The price of energy storage and extra photovoltaic array increases with size; therefore, having too much energy and storage capacity will waste money. We are faced with a trade-off between keeping the costs down and meeting the energy demand during low-solar-energy periods.

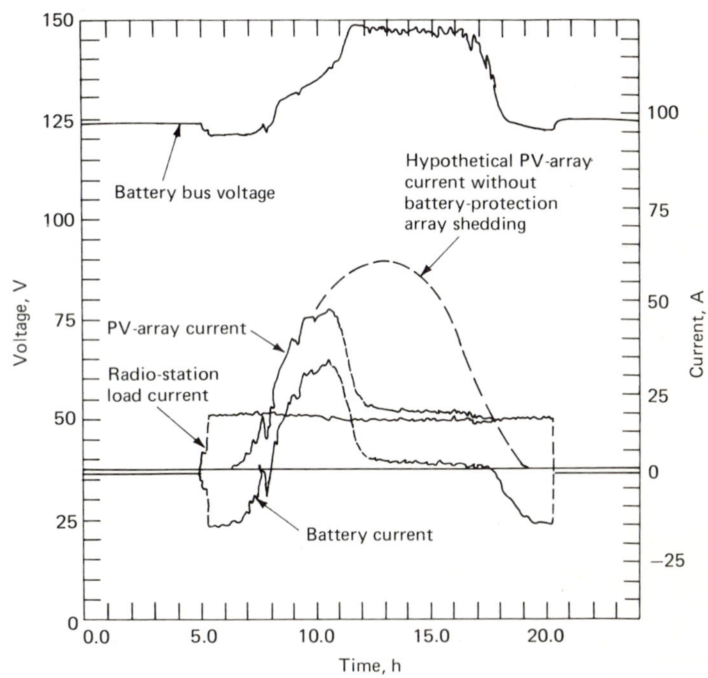

Figure 8-1 **Current flow in a battery-charging photovoltaic system. (Ohio Radio Station, M.I.T. Lincoln Laboratory.)**

Putting the solar-energy surplus to work performing various tasks that are switched in when there is excess energy will decrease waste. Another solution would be to undersize the storage capacity and array and to meet periods of deficiency with auxiliary power, such as the utility or a diesel generator. On the other hand, the load may have to diminish or cease its energy demand during brief periods when solar energy is low.

Utility Storage

The utility buyback rate will be a crucial factor in sizing a photovoltaic array in a utility-interactive system. If the buyback rate is high, it might be beneficial to oversize the array and sell extra energy back to the utility. If the rate is low, the designer may be well advised to undersize the array and sell back little. In the latter case, the utility would serve as a backup power supply rather than as storage during low-insolation months. The ratio between total and directly used photovoltaic energy as measured for a residential photovoltaic system in the northeastern United States is depicted in Figure 8-2 (these data were measured at the Northeastern Residential Experiment Station, described in Appendix A). Notice in this example that between one-fifth and one-half of the total photovoltaic energy generated is directly utilized. A significant portion of the energy

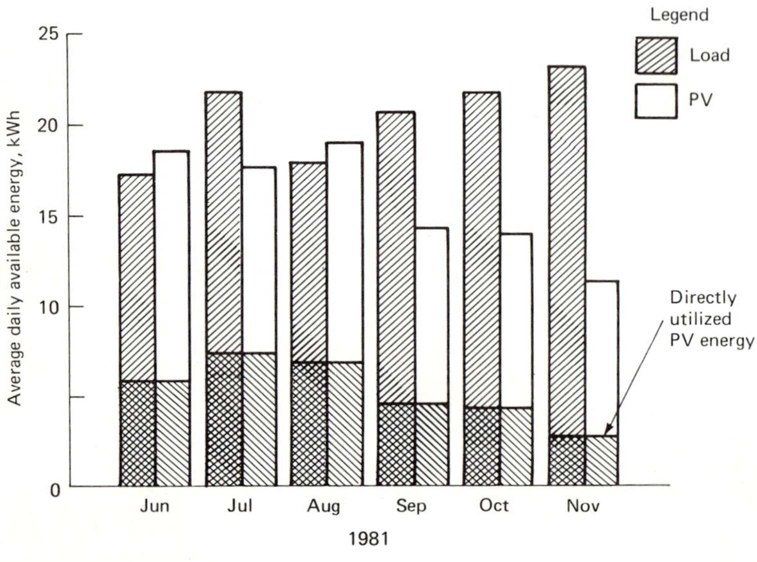

Figure 8-2 Energy transfer between Trisolar PV prototype, monitored residence number 2, and utility. (Northeast Residential Experiment Station, M.I.T. Lincoln Laboratory.)

demanded by the residential load during the daytime hours cannot be provided by the PV array, as the load power demand often comes in brief peaks that sometimes exceed the solar power available (as is evident in Figure 7-1); partly cloudy or totally overcast conditions also cause photovoltaic power to be less than needed. These facts underscore the need for a favorable buyback ratio in order to provide the PV system owner with a good price for the energy that invariably will have to be sold back. These important cost-related issues are discussed further in a later section.

System Design Sample Problem

A family located in Concord, Massachusetts, wants to design a photovoltaic system to meet their electric energy needs. Their residence uses a combination of gas and solar for heating in the winter and gets by without any high-energy-consumptive air-conditioned cooling in the summer. Their electricity demand is for basic appliances needed for lighting, cooking, cleaning, refrigeration, and entertainment. They have an 80-m^2 south-facing roof with a 45° tilt which has good exposure to sunshine throughout the day and year. They want to know how large an array will be needed to meet their energy demand and therefore proceed to follow the step-by-step design procedure.

1. The monthly average daily energy demand for each month of the year and a graphic plot of energy output versus time of year are found in Figure 8-3.
2. The array will be sized to satisfy the annual average daily energy demand of 15.9 kWh/day.
3. The insolation data for Boston, Massachusetts, from Appendix C and the conversion equations to adjust horizontal to 45° tilted insolation are used to predict the solar energy at the site; these are compiled in Table 8-1.
4. The array will be sized according to the annual average solar energy for Boston: 3.9 kWh/m^2 · day.
5. The photovoltaic system will be utility-interactive and will use a fixed voltage input line-commutating inverter with an average efficiency of 90% (utility buyback ratio = 1, with no time constraints).
6. Solar-cell modules available on the market are chosen that have a module efficiency at 20°C of 11.28% (which accounts for cell packing factor = 80%). The annual average daytime temperature, deter-

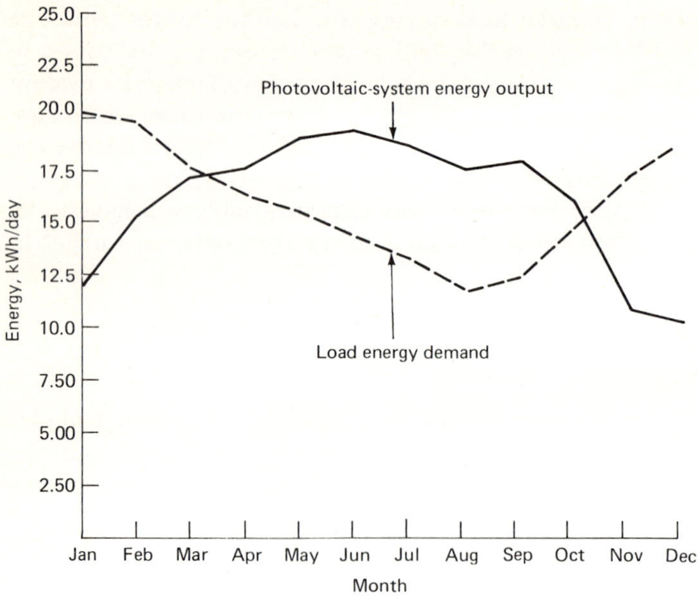

	Residential-load energy input, kWh/d	PV-system energy output, kWh/d		Residential-load energy input kWh/d	PV-system energy output, kWh/d
Jan	20.1	12.0	Jul	13.3	18.7
Feb	19.2	15.2	Aug	11.8	17.4
Mar	17.6	17.0	Sep	12.5	17.9
Apr	16.2	17.6	Oct	14.6	15.9
May	15.5	18.9	Nov	17.1	11.0
Jun	14.3	19.2	Dec	18.9	10.2

Figure 8-3 **Sample problem monthly average daily energy demand and PV-array energy output versus time of year.**

mined by averaging the monthly figures for Boston found in Appendix C, equals 11.4°C. The cell temperature is about 25°C higher on the average, and therefore will be about 37°C (the nominal operating cell temperature of the modules when they are mounted in an architecture that resembles the final design is about 45°C). The temperature correction factor at a cell temperature of 45°C taken

Table 8-1 Sample Problem Calculations for Converting Solar Energy on a Horizontal Surface to Solar Energy on a Tilted Surface

LOCATION: Concord, Massachusetts, U.S.A
LATITUDE: 42.45° North
SURFACE TILT ANGLE: 45° facing due South

Month	δ, degrees	$\overline{H}$ kWh/m² • day	$\overline{K}_T$	ω_s degrees	$\overline{H}_d/\overline{H}$	ρ	$\overline{R}_b$	$\overline{R}$ kWh/m² • day
JAN	−20.9	1.50	0.396	69.6	0.470	0.7	2.530	2.77
FEB	−13.0	2.24	0.425	77.8	0.439	0.7	1.940	3.51
MAR	− 2.4	3.21	0.435	87.8	0.429	0.4	1.440	4.01
APR	9.4	4.18	0.445	89.6	0.419	0.2	1.080	4.25
MAY	18.8	5.26	0.482	89.1	0.386	0.2	0.891	4.76
JUN	23.1	5.73	0.498	88.9	0.372	0.2	0.818	4.93
JUL	21.2	5.52	0.493	89.0	0.377	0.2	0.849	4.86
AUG	13.5	4.69	0.471	89.4	0.396	0.2	0.992	4.53
SEP	2.2	3.97	0.488	89.9	0.381	0.2	1.280	4.55
OCT	− 9.6	2.81	0.469	81.1	0.397	0.2	1.760	4.01
NOV	−18.9	1.59	0.377	71.7	0.491	0.2	2.350	2.62
DEC	−23.0	1.27	0.381	67.2	0.487	0.4	2.740	2.39

from Figure 4-4, is about 0.9. Using Equation (8-2), we calculate the array area to be about

$$A = \frac{15.9}{4.1 \times 0.1128 \times 0.9 \times 0.9 \times 0.9} = 47.2 \text{ m}^2$$

7. The estimated array peak power is found using Equation (5-2).

$$\text{PP} = 47.2 \times 1000 \text{ W/m}^2 \times 0.145 \times 0.7 = 4790 \text{ Wp}$$

8. The utility-interactive inverter input-voltage range is between 150 and 220 V DC. The array maximum-power voltage should be about 180 V to allow leeway in either direction for temperature-induced voltage migrations. The array area should be about 50 m², and the array peak power is about 5 kWp. The modules that have been chosen have a maximum-power voltage of about 10 V with a peak power of 135 Wp and an area of 1.33 m². We therefore decide to build the array out of two series strings of 18 modules each wired in parallel

branches; 36 modules are thereby wired into a configuration that produces 180 V. The array area will therefore be 47.8 m², and the peak power rating at 20°C will be around 4.86 kWp.

9. The monthly average daily energy output for our proposed photovoltaic system is roughly predicted using the solar-energy data for a tilted surface calculated in step 3 and Equations (5-1) and (8-1). The projected photovoltaic energy output for an average day of each month is listed in Table 8-2 and plotted in Figure 8-3.

10. Comparing the load energy demand and photovoltaic energy supply curves, we notice that their respective peaks do not occur at the same time of the year; the photovoltaic array generates the most electricity

Table 8-2 Sample Problem Photovoltaic-System Calculations

MODULE EFFICIENCY: 11.28% @ 20°C
MODULE PACKING FACTOR: 0.9
ARRAY AREA: 47.8 m²
SOILING FACTOR: 1.0
POWER CONDITIONING EFFICIENCY: 90%

Month	Average tilted energy, kWh/m² · day	Average daytime cell temp,* °C	Temp. correct. factor†	Average PV energy, kWh/day	Average system energy, kWh/day
Jan	2.77	23.9	.99	13.3	12.0
Feb	3.51	24.7	.99	16.9	15.2
Mar	4.01	28.5	.97	18.9	17.0
Apr	4.25	34.3	.95	19.6	17.6
May	4.76	40.1	.91	21.0	18.9
Jun	4.93	45.4	.89	21.3	19.2
Jul	4.86	48.5	.88	20.7	18.7
Aug	4.53	47.6	.88	19.3	17.4
Sep	4.55	43.8	.90	19.9	17.9
Oct	4.01	38.7	.91	17.7	15.9
Nov	2.62	32.8	.96	12.2	11.0
Dec	2.39	26.3	.98	11.4	10.2

*Approximated by adding 25°C to the adverage daytime temperature at the Boston site as found in the Appendix.

†As determined from the graph in figure 4-4

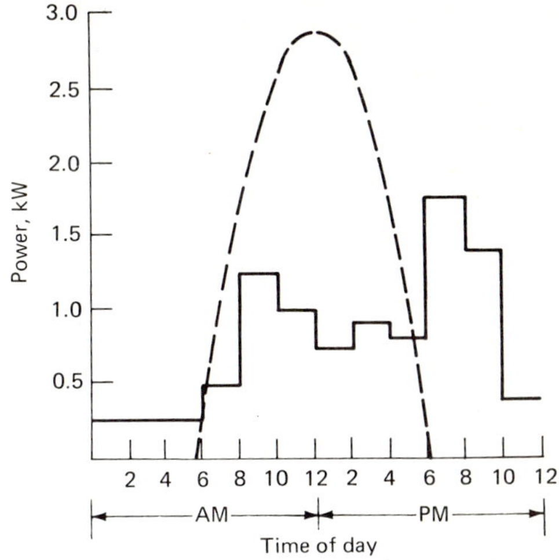

Figure 8-4 Sample problem power demand and PV-array power output versus time of day.

in the summer, when the load demand is at its lowest. Long-term energy storage clearly will be needed to shift the excess solar-cell energy in the summer to meet the excess load energy demand in the winter. A utility-interactive inverter will enable the residence to use the utility grid as storage with an annual net energy flow of zero.

11. A graph that approximates the electric energy demand of the residence over a typical 24-h period during the spring is shown in Figure 8-4.

Figure 8-5 Trisolar prototype. (Northeast Residential Experiment Station, M.I.T. Lincoln Laboratory.)

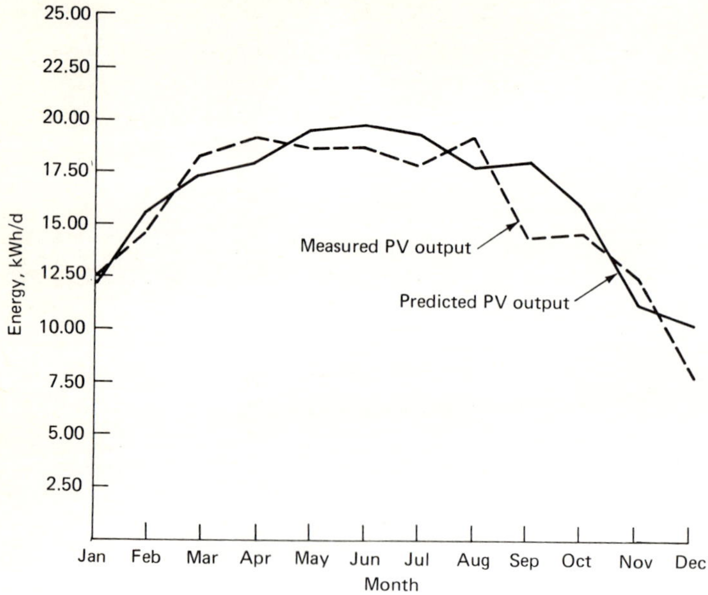

Calculated PV Energy, kWh/d	
Jan	12.0
Feb	15.2
Mar	17.0
Apr	17.6
May	18.9
Jun	19.2
Jul	18.7
Aug	17.4
Sep	17.9
Oct	15.9
Nov	11.0
Dec	10.2
Avg. daily energy	15.9

Measured PV Energy, kWh/d	
Jan	12.4
Feb	14.2
Mar	18.0
Apr	18.8
May	18.3
Jun	18.4
Jul	17.6
Aug	18.9
Sep	14.2
Oct	14.6
Nov	12.3
Dec	7.6
Avg. daily energy	15.4

Figure 8-6 Predicted and measured PV-array energy output versus time of year, measured between May 1981 and April 1982. (Northeast Residential Experiment Station, M.I.T. Lincoln Laboratory.)

12. The approximate average peak array power during the spring is equal to the array peak power of 4.79 kWp multiplied by the annual average percentage of possible sunshine for Boston of 59% (as found in Table 3-2). In the spring the sun rises at about 6 AM and sets at about 6 PM. Figure 8-4 shows the plot of solar power versus time of day that is extrapolated from the three data points established above.

13. The PV output-power curve peaks around the middle of the day, while the load input power peaks a few hours after sunset. Short-term storage will be needed to transfer excess daytime solar-cell energy to meet the excess load demand in the evening. The utility-interactive inverter will perform this function well.

A residential photovoltaic system with the general specifications outlined in this sample problem has been constructed at the Northeast Photovoltaic Residential Experiment Station in Concord, Massachusetts, as shown in Figure 8-5. The predicted PV-system output-energy/time-of-year curve is compared with the actual measured output energy for one year in Figure 8-6. The two curves show a correlation that is as good as can be expected given the unpredictable elements in weather forecasting.

PHOTOVOLTAIC-SYSTEM COSTS VERSUS BENEFITS

There is no doubt that photovoltaic power is capable of meeting most electricity needs; the real issue is whether photovoltaic electricity is competitive with other energy-generating options in terms of benefits versus costs. In contrast with most conventional electric-power plants, the fuel that photovoltaic generators use is free; hence the operating costs are minimal. However, the initial investment or capital cost on a PV system is large and must be measured against a payoff in the form of energy cost savings. The acquisition of a photovoltaic electric generator involves buying solar-cell modules, power-conditioning equipment, wire, and miscellaneous interconnect and safety devices. It also entails paying for design and installation labor, and for such operating expenses as maintenance, property tax, and insurance. Gradually, this power source produces a return on investment consisting of the electricity cost savings that are accumulated over the system's lifetime by not buying electricity from other established sources. The costs of a photovoltaic power installation must be measured against its long-term gains.

The economic feasibility of a photovoltaic system can be assessed by subtracting the present value of total costs from total financial benefits. Weighing benefits against costs is complicated by various changing factors that must be evaluated over the system's lifetime. The prevailing price of electricity in particular, the price of energy and goods in general, and the discount rate for borrowing money or for gains on alternative investment opportunities, will vary with time. An educated guess must be made as to how these critical variables will change, and a common reference point in time must be chosen to make the benefit-versus-cost comparison.

A commonly used economic evaluation technique called *present-value analysis* will enable us to measure cost-effectiveness by normalizing all financial transactions to the present.* All the costs and benefits that are affected by money's changing value over time are adjusted to the present by multiplying them by a uniform present worth factor that accounts for life-cycle price escalation and loan or investment discount rates. If the net present value (the benefits minus the costs) is positive, the solar electric system is economically viable. A negative present value indicates that the system will probably not pay off and that certain design features or costs will have to change. This economic study is only as valid as its underlying financial predictions and design information.

There are important benefits associated with photovoltaic energy systems that will not be addressed in this economic evaluation because they are too difficult to price: the advantage of energy independence; the increasing importance of generating energy without environmental or noise pollution; the benefit of a technology that does not require a high degree of sophistication to understand, construct, and maintain; and the social and economic value of an industry that provides many and varied jobs. An assessment of these benefits is based on largely subjective considerations that are beyond the scope of this work. Nonetheless, it is recognized that a PV system may be found to have a negative net present value and yet be attractive because of other favorable features.

System Costs

Appraising the cost of establishing and operating a PV system will involve itemizing the various components and procedures involved. Almost all the costs are initial and pertain to equipment and labor at the time of design and installation. This large purchase price can be deferred by taking out a loan, but borrowing money also has its price—especially during hard economic times. The first step in a cost evaluation will be to make a systematic list of all the expenses associated with photovoltaic electricity.

The price of a photovoltaic array is deduced from the price per peak watt that is typically quoted for photovoltaic modules. Once array area, packing factor, and system efficiency are known, the PV array's peak power can be calculated by using Equation 5-3; the PV array's price then simply equals the total peak power multiplied by the existing price per peak watt. (*Note:* the temperature at which peak wattage and array efficiency are measured must be identical and as close to NOCT as pos-

*Other economic evaluation methods, with varying advantages and disadvantages, exist but will not be discussed here.

sible.) The support structure and mounting design for the solar-cell modules must be chosen, and a financial quote on the materials involved should be obtained. If the array is mounted directly or integrally on a roof, credit must be deducted for the roofing materials that are displaced.

The cost of power-conditioning equipment, whether a DC-to-AC inverter, an energy storage system, or a maximum-power tracker, must be established for units that satisfactorily meet the size requirements under consideration. Once we know the array size and location, we can approximate the length and type of wire and the numberof connectors, switches, terminals, safety devices, and other such components that will be needed; general prices must then be obtained. A ground-mounted array will have expenses for land and possibly a fence. Power-conditioning equipment may need to be housed in a special room or building, which also will increase the overall price. The labor-hours and labor charges needed to design and install the PV system completely can be estimated by an electrical engineer and/or contractor.

There will be small operating charges during the system's lifetime. Maintenance will involve washing the array periodically and repairing any problems that might occur. The annual property tax on land or a house that has a photovoltaic array on it will probably increase. The size of such an increase can be assessed by local tax officials. It generally will be wise to insure the photovoltaic system (especially if it is located on a roof where damage to the house could occur); the annual insurance costs will be another expense.

Since maintenance, property tax, and insurance premiums will accrue over the system's lifetime, they will be affected by the general rate of price inflation and the discount rate for borrowing money. Money used in the future will have a different value today. At this point, an extended prediction must be made of how fast prices will rise and how much it will cost to borrow money. As will be discussed in the following section, this prediction is based on both objective and subjective interpretations of economic events. The three operating expenses are normalized to the present using the uniform present worth factor found in the tables of Appendix D; to look up the UPW factor, it is necessary to know the system's expected lifetime and the predicted inflation and loan discount rates throughout this period.

There are various tax incentives that will allow private citizens and companies to deduct part of the cost of buying a solar energy system. These incentives change with the political climate and vary from one district to the next. An individual will often be able to deduct a significant part of the overall system cost from his or her federal and state income tax. Businesses can depreciate a photovoltaic system as its value declines over its productive lifetime, and thereby reduce the system cost. These tax options for saving money on the purchase of a solar electric generator need to be examined and taken into account.

After calculating the present value of the system's total cost (as detailed later in the cost-versus-benefit analysis procedure), the designer may find the price too high to pay all at one time, and therefore will need to take out a loan. Borrowing money, however, will increase the system's final cost. After a down payment has been made on the loan, the remaining sum (principal) will be financed over a prescribed time period (term) at the current discount (interest) rate. Based on this information, the monthly payments can be calculated and the total cost of the loan (interest) can be ascertained. (*Note:* Interest can be tax deductible). Since monthly payments will be made over an extended period of time, during which inflation will erode the value of money, it will be necessary to use the appropriate uniform present worth factor to compute the present value of the loan payments (in this case, the UPW factor is based on the loan's term, the discount rate, and the average predicted inflation rate throughout the term). Note that an increase in the loan discount rate will raise the cost of borrowing money and the final cost of a PV system; a healthy economy encourages investments and vice versa.

At the end of the PV system's useful life, much of the hardware will have a salvage value (especially in the coming decades, as resources become more scarce). Many of the structural and electrical components can be reused or recycled. Making a precise prediction of the salvage value will be impossible, but 5% to 15% of the present price of the solar-cell module, array support structure, power-conditioning, and wiring material costs will serve as an estimate.

System Benefits

The financial benefit from a PV system will be the relative value of its electric output. In other words, if we were to sell the electricity produced by a proposed photovoltaic system as it is generated, how much would we earn over the system's lifetime? To evaluate this benefit, we must first determine how much energy our solar electric generator will produce. Part of this electricity, unfortunately, may never be put to use; some types of systems will be forced to discard or undersell energy surpluses. Therefore, we may need to estimate a utilization factor that will indicate how much electricity will actually be applied. We need to know the price of electricity from other available sources, and to predict how this price will change during the PV system's period of operation. Our interpretation of world events will play a key role in this cost-increase prognosis. Other investment opportunities will also be attracting our money, and we may want to make a direct comparison between two options. Present-worth analysis will enable us to take inflation and the gains on another investment strategy into account as we normalize all the money earned to the present.

To predict the annual energy output of our proposed PV system, we will need to know the array's area and packing factor and the system's efficiency (at NOCT). The annual average daily solar energy available on an appropriately tilted south-facing surface at the site can be ascertained using the data available in Appendix C or Figure 3-10 and/or the equations found in Chapter 3. The insolation data are in the form of energy per monthly average day; determining the annual average daily insolation will involve averaging the twelve monthly figures and multiplying by the number of months in a year (12). The energy output of a given PV system is calculated using the equations found in Chapters 5 and 8.

Certain systems, primarily those involving some form of storage, will be forced to squander or undersell some of the PV system's energy output. For instance, battery storage is usually sized to meet the system's average charge/discharge rate, but during certain peak insolation periods the solar electric output may exceed the battery's storage capacity. Excess energy will have to be diverted or discarded. In a utility-interactive system, the utility may decide to buy excess electricity from a photovoltaic cogenerator at less than the normal selling rate. In other words, the buyback ratio (PV/utility electricity selling ratio) may be less than 1. Under such circumstances, the surplus electricity generated by the PV array during the day will be worth less than the electricity used directly. The *utilization fraction* is an indication of how much solar electricity is discarded or undervalued. This fraction is difficult to gauge because it is dependent upon such changing variables as weather patterns, electricity demand, and energy prices. The utilization fraction can be approximated using the power versus time of day or the energy versus time of year plot that was formulated for the load and the proposed PV system during the system-sizing procedure (described earlier in this chapter). From these plots, we can determine the amount of excess energy that is generated by the PV system during a typical day or throughout the year.

When using short-term storage, such as a battery, most surplus energy during the year will have to be diverted or discarded; therefore, the energy versus time of year plot will give us a better indication of the short-term storage utilization fraction. With unlimited storage, such as a utility-interactive inverter, the surplus energy generated during each day may be valued at less than the energy used directly; the power versus time of day plot will give us a better indication of the long-term storage utilization factor. If the buyback ratio established by the local utility company is equal to 1 (the cost of electricity sold to the utility equals the cost of electricity bought), then the utilization fraction also equals 1. The utility may, however, make the buyback ratio conditional upon different factors that could complicate the issue (for example, if the buyback ratio could be valid only within a limited time period, such as 1 month).

Once careful plots of the expected PV output and load input throughout the day and year have been made, the utilization fraction can be estimated using Equation (8-3).

$$\text{UF} = \frac{\text{TPO} - \text{SPO} \times (1 - \text{BBR})}{\text{TPO}} \tag{8-3}$$

where UF = utilzation fraction (0 to 1)

TPO = total PV output (kWh/day or kWh/year)

SPO = surplus PV output (kWh/day or kWh/year)

BBR = buyback ratio (0 to 1)

The total PV output for the year or for an annual average day can be obtained from the information developed during the system-sizing procedure. The surplus PV output must be extrapolated from one of the two PV/load output/input plots as shown in Figure 8-7. The area (measured in kilowatthours) between the PV load curves when the PV output exceeds the load input must be approximated. This determination will sometimes tend to overestimate the utilization fraction, since the actual load input curve, as shown in Figure 7-1, for example, and the actual solar-energy plot, as shown in Figure 3-13, will be more erratic than our approximated curves indicate. The sellback ratio for a system involving battery storage is equal to zero (since all the energy is retained or discarded), while the sellback ratio for a utility-interactive system will be established by the local utility and may range anywhere from about 0.5 to 1.

The value of earnings from our proposed solar electric generator, throughout its operation, will change with time when compared with the possible earnings from an alternative investment opportunity or the escalating price of energy. Therefore we will use an appropriate uniform present worth factor to adjust all money to the present. Choosing the right UPW factor will once more require us to make some sensitive economic predictions. Changes of a few percentage points one way or the other in the rate of energy price escalation rate or the returns on alternative investments will have a dramatic impact on the PV system's net present value. A solar electric generator becomes more attractive as the escalation rate of energy prices increases, but conversely becomes less attractive when the returns on alternative investment options appear greater.

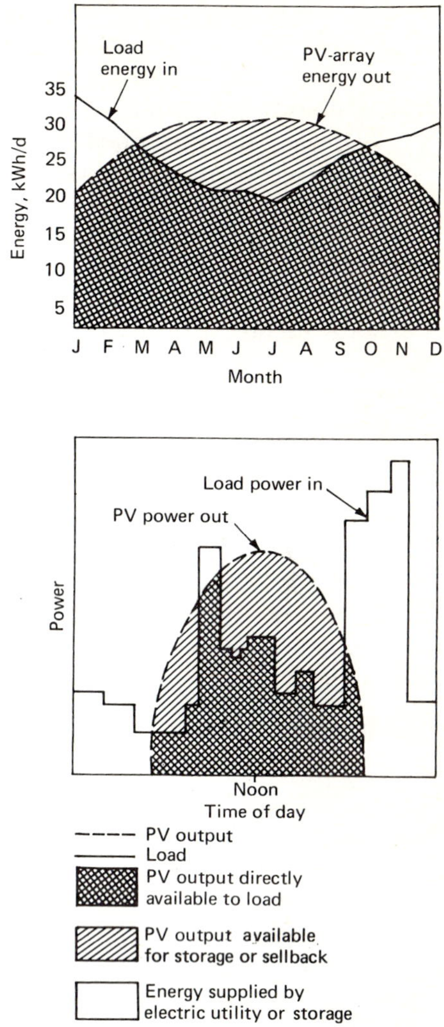

Figure 8-7 Determining the utilization fraction from energy/time-of-year and power/time-of-day plots.

When deciding how to spend our money, we will be faced with various options. Money can be put in a savings account or in stocks and bonds, where it will grow at the prevailing interest rate. Other energy systems may show promising returns. In any case, a realistic comparison should be made between a photovoltaic system and an alternative investment scenario. The payback period of an investment (the time it takes to recoup one's investment) is a commonly used piece of information that enables one to make simple comparisons. The time needed for payback can be approximated using present-worth analysis by changing the "expected lifetime" of a venture until its net present value just turns positive; the "expected lifetime" then equals the payback period.

We will also want to compare the value of our photovoltaic electricity with that generated by the next most competitive source. The generally available price of electricity will most likely increase at a continuous pace over the coming decades. For instance, average electricity prices in the United States rose at an annual rate of 13.1% between 1976 and 1981 (which is higher than the consumer price inflation rate of 11.0% over that same period). Inflation rates may vary from region to region depending on the local availability of energy and other goods and services.

There is no reliable way of predicting how fast energy prices or investment gains will increase in the future. If we base our forecast on the rate of previous years, then we assume that the old pattern will continue. Even a detailed study of economic trends will require many assumptions. Although there is no dependable way of making economic forecasts, they will play a major role in the assessment of economic viability. At this point, the designer must make an educated, as well as intuitive, supposition concerning the world's economic outlook.

Net Present Value

Having independently assessed the present value of the benefits and costs for a proposed PV system, we are prepared to calculate its net present value by subracting total costs from total benefits. A positive net present value indicates that the system will pay for itself within its lifetime, assuming that all the criteria are correct. A negative net present value, however, demonstrates that one or more unfavorable conditions exist: some of the system components are too expensive; the current interest rate is exorbitantly high; the tax incentives are not favorable; the proposed site does not receive sufficient sunlight; local electricity prices are low and are expected to remain so for the next few decades, and so forth. In order for the photovoltaic system to be cost-effective, certain design and/or cost citeria will have to change.

Most of the economic and political factors that affect the cost of a photovoltaic system are beyond the individual designer's control. It is, nonetheless, possible to optimize the PV array size for the load energy demand in a way that achieves the most cost-effective match. Given the established annual energy demand profile, a change in the array size (and thus in PV output power) will also change the utilization fraction. Stated differently, if the amount of energy used remains constant, but

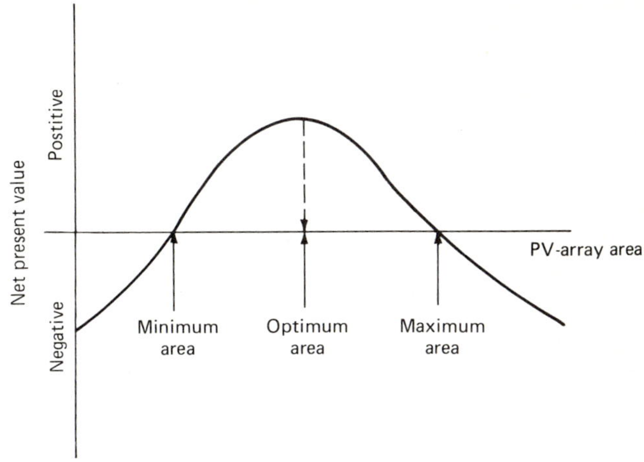

Figure 8-8 Net present value of PV system versus PV array area.

the quantity of energy generated changes, the amount of energy that is effectively utilized will also change. If the change in utilization fraction that is produced by a change in array area can be calculated to a reasonable degree of accuracy, and if an estimate can be made of how such size-variable costs as array structure, wiring, power conditioning, and labor increase with array size, a plot of net present value versus array area can be made. Such a graphical representation can demonstrate the possible range of array sizes that will have a positive net present value, and also indicate what the optimum or most economical array size will be. A sample of such a plot is shown in Figure 8-8. A graph of net present value versus array area depends on reliable data and is greatly facilitated by using a computer.

The following sample problem presents a basic procedure for evaluating the cost-effectiveness of a photovoltaic system. In this example we will examine all the costs and benefits associated with establishing a utility-interactive photovoltaic system in Orlando, Florida (latitude 28.5°). The photovoltaic array is positioned due south at a 30° tilt angle from the horizontal. The system uses a 90% efficient inverter and is able to sell electricity back to the utility at a buyback ratio of 100%. The present price of electricity is $0.10/kWh, and the expected rate of electricity price escalation over the 20-year lifetime is 15% (between 1976 and 1981, electricity prices in the United States rose at an annual rate of 13.1%). The PV array cost is $2.50 per peak watt (in 1981, PV modules were selling for $7.00 to $12.00 per peak watt). Here we are assuming that the price of PV modules will come down, as is generally expected.

SAMPLE PROBLEM: Present Value of Photovoltaic System Benefits versus Costs

PHOTOVOLTAIC SYSTEM CRITERIA:

1.	PV-array size A, m^2	50
2.	PV-array packing factor PF (a fraction between 0 and 1)	0.75
3.	PV module efficiency PVE at NOCT, %	10
4.	Power-conditioning efficiency PCE, %	90
5.	PV-array peak power PP at NOCT, Wp PP = 1000 × A × PF × PVE/100	3,750
6.	PV-system expected useful lifetime L, yr	20

PHOTOVOLTAIC SYSTEM COSTS:

1.	Peak-watt PV module costs MC, $/Wp	2.50
2.	PV-array costs PVC, $ PVC = MC × PP =	9,375
3.	PV-array support-structure costs SC, $	2,500
4.	PV-system power-conditioning costs PC, $	2,750
5.	PV wiring-materials costs WC, $	1,000
6.	Design and installation labor costs LC, $	2,750
7.	Annual property tax increase PT resulting from addition of PV system, $	50
8.	Annual insurance premium for PV system IC, $	100
9.	Annual PV-system maintenance costs MC, $	100
10.	Predicted average annual general price escalation rate over lifetime of PV system ER1, %	12
11.	Predicted average annual discount rate for borrowing money over lifetime of PV system DRI, %	15
12.	Uniform present worth of costs UPW1 based on L, ER1, and DR1 (see Appendix D)	15.33
13.	Percentage of PV-system costs deducted for tax or depreciation credit TC, %	30
14.	Salvage value SV after system lifetime, $	1,000
15.	Total PV-system costs TPC, $ TPC = {[(PP × PVC) + SC + PC + WC + LC + UPW1 × (PT + IC + MC)] − SV} × (100 − TC)/100	14,545.25
16.	Loan down payment DP, $	5,000
17.	Term of loan TL, yr	10

18. Loan discount rate DR2, % — 15

19. Predicted average annual general price escalation rate over term of loan ER2, % — 12

20. Loan uniform present worth UPW2 based on TL, ER2, and DR2 (see Appendix D) — 8.67

21. Loan monthly payments MP* throughout duration of term, $

$$MP = \frac{(TPC - DP) \times [(DR2(1 + DR2)^{TL}]}{12 \times [(1 + DR2)^{TL} - 1]} =$$ 158.49

22. Present value of PV system-costs NSC, $

NSC = DP + (TPC − DP) × (TL/UPW2) = 16,009.50

PHOTOVOLTAIC SYSTEM BENEFITS:

1. Annual average daily solar energy SE, kWh/m^2·d — 4.9
2. Utilization fraction UF (a fraction from 0 to 1) — 1.0
3. Present available electricity costs EC, $/kWh — 0.10
4. Predicted average annual escalation rate of electricity prices over lifetime of PV system ER3, % — 15
5. Discount rate for alternative investment opportunities DR3, % — 12
6. Benefit uniform present worth UPW3 based on *L*, ER3, and DR3 (see Appendix D) — 26.71
7. Annual PV-system useful output PVO, kWh/yr

   PVO = 365 × SE × A × PF × PVE/100 × PCE/100 × U = 6,036.2
8. Present value of PV-system output over lifetime VPO, $

   VPO = PVO × EC × UPW3 = 16,122.66

NET PRESENT VALUE:

1. System benefits minus costs NPV, $

   NPV = VPO − NSC = +113.14

*The interest on a loan can be tax-deductible; this benefit has not been included in these calculations.

9

PHOTOVOLTAIC ARRAY INSTALLATION

The installation of a photovoltaic array can be divided into two separate tasks: physical and electrical. The physical installation involves mounting the solar-cell modules on a sturdy frame that will maintain the array's tilt and position throughout the worst weather conditions. The electrical installation entails wiring the modules together in their prescribed parallel and/or series configuration and leading the main power lines to the power-conditioning equipment and the load.

PHYSICAL INSTALLATION

A variety of factors must be considered when physically installing a photovoltaic array, especially when the modules are to be mounted on a building. A superstructure must be able to support the modules through adverse wind and weather for the 20-year expected lifetime of the array, and yet be situated in an unsheltered location where sunlight is not blocked. Photovoltaic modules should be installed in a fashion that allows easy access for repair and maintenance. Solar cells operate more efficiently at lower temperatures; consequently, the modules should be installed in a way that will keep them as cool as possible during daytime operation. Roof-mounted photovoltaic modules must maintain the roof's weather seal. The price of a mounting structure must be kept as low as possible. Finally, since photovoltaic arrays often will be installed in areas of high visibility, they need to be physically attractive, or at least should blend in with the environment.

Photovoltaic-array repair during the 20-year projected lifespan of a PV system will entail replacing modules whenever a failure occurs. Large portions of an array can malfunction when an individual module fails; it will be necessary to remove modules for inspection and replacement when module defects occur. If the hardware that attaches the modules to their support structure is designed to allow for easy removal, the task of troubleshooting and repair will be made that much simpler.

Maintenance of a photovoltaic array will include periodically cleaning the module's cover glass. In the winter time, snow can obstruct most of the sunlight. PV modules, however, tend to naturally shed snow; some sunlight usually penetrates through a snow blanket and heats up a module's dark surface. The snow then melts and slides off a module's normally smooth cover. The speed with which snow shedding occurs will depend on the tilt angle of the array, the amount of thermal insulation modules have on their back surface, and the amount of snow and sunlight. In most cases, an operator will not need to brush snow off an array, since it will take care of itself with little delay. Airborne dirt poses a more serious problem; dirt will soil the cover glass of a photovoltaic module, so that the light intensity reaching the solar cells is reduced and the current and power output are decreased. The amount and type of filth depends on the site's location, typically increasing as the area becomes more urban. Some dirt will wash away with the natural cleansing action of rain and snow, while other dirt can be washed away only with detergent and a scrub brush. A photovoltaic-array mounting design should consider how array washing will be performed.

The energy conversion efficiency of a silicon cell increases with decreasing temperature at a typical rate of about 0.5% 1°C (relative) of the 25°C value (i.e., a cell with a 10% efficiency at 25°C will have an efficiency of 9% at 45°C). Keeping the photovoltaic-array operating temperature as low as possible optimizes the amount of energy obtained from the array and increases PV module lifetime. The daytime temperature of a solar-cell module is governed by the insolation, the ambient temperature, and the amount of ventilation or heat transfer available. The last factor is the only one that is somewhat under our control and depends on what kind of environment the solar cells are placed in. If, for instance, photovoltaic modules are mounted in such a way that there is ventilation on both their front and back, they will operate as much as 20°C cooler than if they are mounted flat on a surface that insulates the heat emitted off the back (See Chapter 4).

A photovoltaic array must be positioned facing south in a place where sunshine will not be blocked. Most sites are not, however, entirely free of shadow-casting obstructions. In the city, neighboring buildings often will cast a shadow on an array during certain times of the day. Trees add beauty to a neighborhood, but also block sunlight. Legal, practical, or aesthetic reasons will often make it impossible to eliminate all obstructions of sunlight. The major objective should be to arrange the array and/or nearby structures in a way that does not block sunlight between the hours of about 7:30 AM and 4:30 PM. Shading time during the early morning or late afternoon is far less critical than shadows cast during the hours when the sun is at its apex in the sky. A photovoltaic array must be generally facing south, but can deviate from due south by as much as 20 to 30° east or west without significant reduction in the total annual energy received.

A roof-mounted array must be installed in a way that maintains a good weather seal. Any scheme for mounting modules on a building will require penetrating or substituting the roof's weather seal in order to anchor the array from frame supports on the building rafters or beams. This must be done in a way that maintains a durable seal during the lifetime of the array. Fire hazards due to arcing in the PV circuit or wind tunneling caused by the PV mounting structure must be prevented.

The cost of a photovoltaic-array frame and support must be as low as possible without sacrificing durability. The structure must be able to withstand the worst snow and wind loading conditions expected at the site. The amount and type of materials used and the installation time should be kept to a minimum to reduce the hardware and labor costs.

Since photovoltaic arrays can be quite large and generally are mounted in highly visible locations, their esthetic appeal can be a critical factor. The shape and color of the cells and the module frames can produce patterns that clash with their environment. Solar cells intentionally are made dark in color to absorb as much solar radiation as possible. When solar cells are mounted on a building, the module frame and the space between cells should be a similar dark color to make the array blend together into a color scheme little different from the roof of a house.

There are four basic photovoltaic-module mounting schemes that have been pioneered by industry and research establishments: rack, standoff, direct, and integral. The last three designs listed specifically represent ways of attaching modules to a house with a south-facing sloped roof. The rack design can be used to mount modules on the ground or on a flat roof.

Rack Mount

Photovoltaic modules installed on a horizontal surface, be it the ground or a flat roof, are mounted on a tilted support frame or rack. There is a wide variety of rack designs made of wood or metal; most manufacturers have rack-mounting frames and hardware that specifically fit their modules. Some rack mount designs afford the option of adjusting the tilt angle by simply loosening or moving a bolt. A rack mount provides easy access to the front and rear of the module for maintenance, testing, and removal. The module operating temperature is as low as possible, since outdoor ventilation of the front and back of the modules provides quick heat transfer. The cost of a rack mount can be relatively high compared with other mounting schemes because of the extra material and labor involved.

Racks situated on the grounds must be mounted on a solid concrete

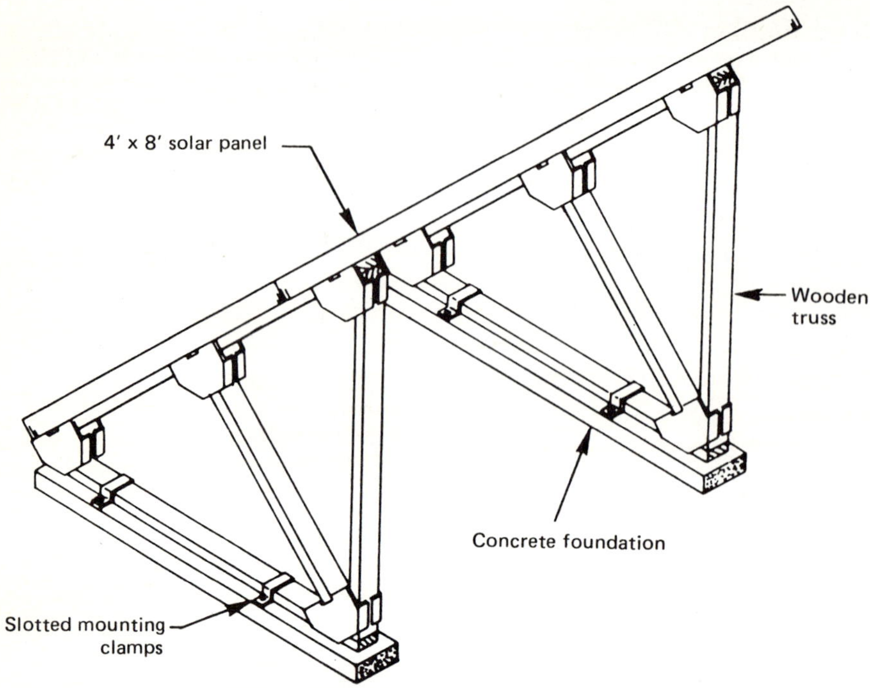

(a)

Figure 9-1 **(*a*) A basic rack ground-mount design. (Y. P. Gupta and S. K. Young, *Design Handbook for Photovoltaic Power Systems,* Sandia Laboratory, Albuquerque, N. Mex., 1981, p. 189.) (*b*) Low-cost rack mount with metal angle bracket frames and anchored concrete block mount. (WBNO radio station, Ohio; photo by M.I.T. Lincoln Laboratory.) (*c*) Required rack-mount spacing to prevent power loss due to shading from one row to next. (Bechtel Group, Inc., *Photovoltaic Subsystem Optimization and Design Tradeoff,* Sandia Laboratory Report 81-0823C, April 1981).**

(b)

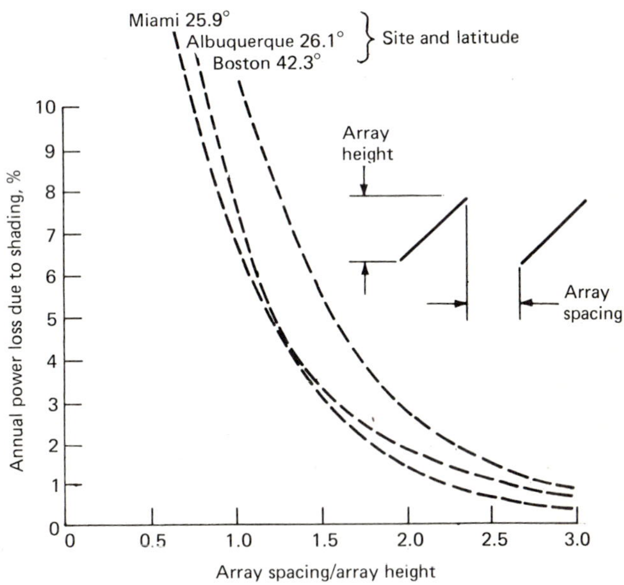
Miami 25.9°
Albuquerque 26.1°
Boston 42.3°
Site and latitude
Array height
Array spacing
Annual power loss due to shading, %
10
9
8
7
6
5
4
3
2
1
0
0
0.5
1.0
1.5
2.0
2.5
3.0
Array spacing/array height

(c)

Figure 9-2 A roof rack mount constructed with wood at SWRES (M.I.T. Lincoln Laboratory.)

foundation to prevent high winds from uprooting the array support. Poured concrete or concrete blocks with auger-type anchors screwed into the ground, as shown in Figure 9-1*a*, will serve this purpose well. A simple metal frame support using angle brackets, as shown in Figure 9-1*b*, provides the superstructure. Large arrays that are mounted on the ground are arranged in rows that are set sufficiently far apart to prevent one row from shading another, as explained in Figure 9-1*c* (the acreage between rows of modules is wasted space that must be taken into account when sizing the area needed to generate a given amount of photovoltaic electricity).

Commercial and industrial establishments and residences in the arid Southwest often have flat roofs. A rack mounted on a flat roof will catch wind and will need to be solidly seated in the roof beams or frame of the house. The wooden rack shown in Figure 9-2 configures the array in one large row that stands on a conventional tarred roof.

Standoff Mount

A standoff roof-mounting scheme, shown in Figure 9-3, involves a module support structure that is mounted over but stands off from a conventional slanted shingle roof. The frame support has to be mounted through the roofing material, requiring careful weather sealing at these points. Modules in a standoff configuration operate at a relatively cool temperature because ventilation is available on both sides of the module. There are some standoff mounting designs that make removal and maintenance of photovoltaic modules quite simple. A "garage door" design

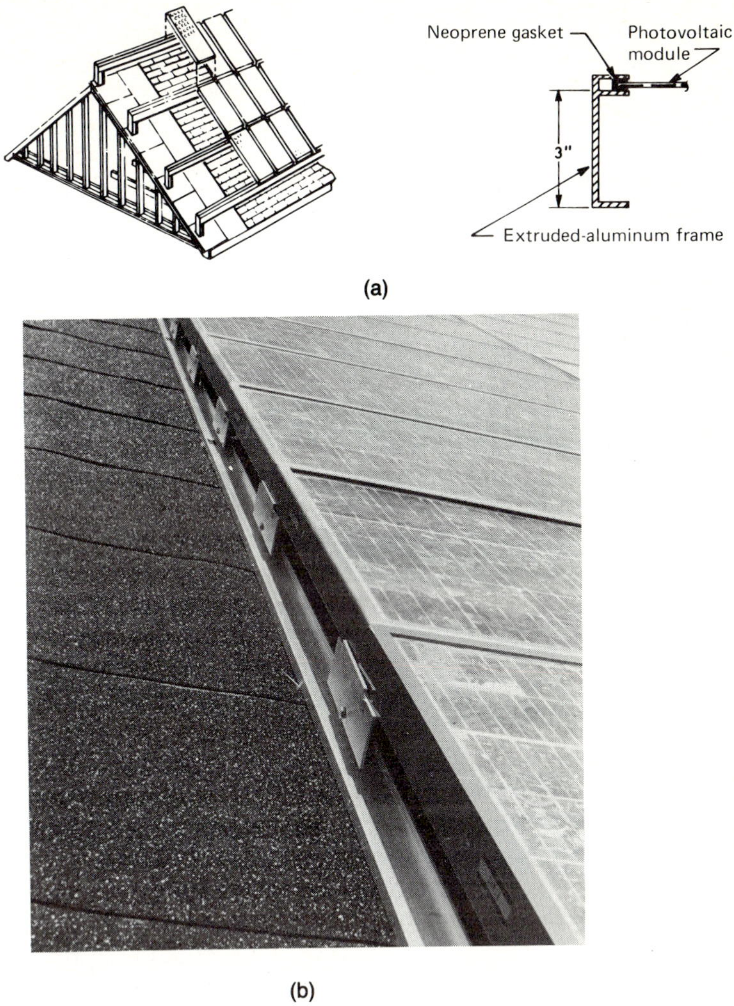

Figure 9-3 (*a*) Basic standoff mount design (M.I.T. Lincoln Laboratory.) (*b*) Polycrystalline solar-cell modules mounted standing off of a conventional shingle roof; modules are mounted with wheels on a track for easy installation and removal. (M.I.T. Lincoln Laboratory prototype, NERES Concord, Mass.; photo by author.)

would enable a person to hoist a string of modules up on a slanted roof on rollers by using a crank and cable, as shown in Figure 9-4. The cost of the standoff approach is higher than that of the following two mounting alternatives because of the sizable amount of hardware involved and the lack of cost savings created by replacing some or all of the roofing

Figure 9-4 **Installing modules on a residential roof in a standoff configuration by rolling them up on the roof with a "garage door" type roller and track mechanism. Carlisle house, Carlisle, Mass. (Photo by M.I.T. Lincoln Laboratory.)**

materials. The standoff mount is an attractive alternative for retrofit applications where the roof already is in place. A standoff mount can produce a wind tunnel that causes fire to spread rapidly; this danger should be considered and avoided.

Direct Mount

The asphalt shingles of a conventional roof can be replaced by solar-cell shingles in a direct-mount configuration, shown in Figure 9-5. Notice how the hexagonal solar-cell modules are mounted directly on the roof felt in an overlapping arrangement that maintains a good weather seal. The sturdy module design and the flush roof mount on a rigid surface make it possible to walk directly on the modules to perform any maintenance. The hexagonal modules are mounted on clips that are anchored into the roof, and are wired together using flat ribbon cables that lie between the module and the roof. The direct mounting of modules on a roof substrate affords no ventilation of the module's back surface, and as a result the solar-cell operating temperature in the daytime is on average about 20°C higher than with the rack, integral, or standoff mounting scheme (NOCT = 60° to 70°C). A higher operating temper-

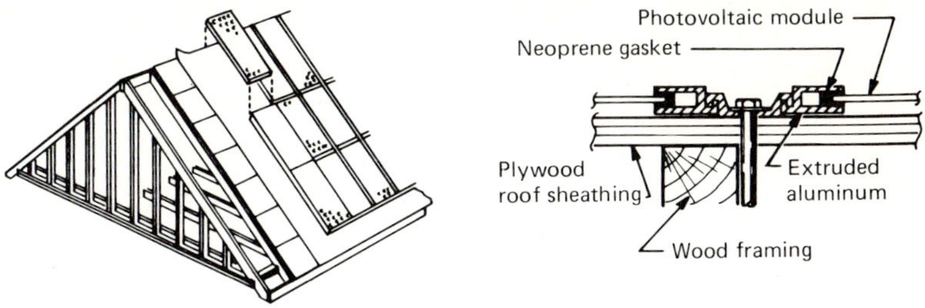

(a)

(b)

Figure 9-5 ***(a)*** **Basic direct-mount design. (M.I.T. Lincoln Laboratory.)** ***(b)*** **Hexagonal shingle modules being mounted directly on the roof felt in an overlapping configuration. (General Electric prototype, NERES Concord, Mass., photo by author.)**

ature decreases the cell efficiency and substantially increases the chances of temperature-induced malfunctions, such as cell cracking or the degradation of materials. Direct-mount installation costs are fairly low, since there is little mounting hardware involved and a cost savings is made by replacing the roof shingles.

Integral Mount

The integral mounting design replaces a slanted roof with photovoltaic modules, as shown in Figure 9-6. Solar-cell modules are mounted directly on the roof rafters and sealed with a gasket or caulking; cracks between modules are further covered with metal flashing. Tolerances for the temperature expansion and contraction of materials must be considered.

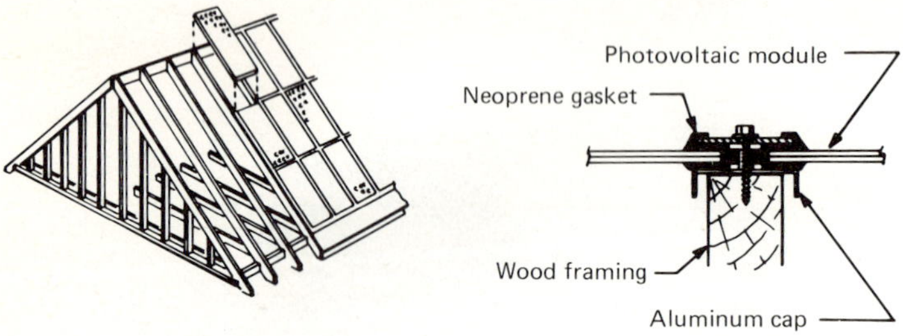

(a)

(b)

Figure 9-6 **(*a*) Basic integral-mount design (M.I.T. Lincoln Laboratory.) (*b*) Integrally mounted photovoltaic modules become the roof of the house. (Trisolar prototype, NERES, Concord, Mass.; photo courtesy of Trisolar Corp., Bedford, Mass.)**

Figure 9-7 Attic view of the back of integrally mounted modules; notice how modules lie directly on the roof rafters without any insulation. (TriSolar prototype at NERES; photo by author.)

A view inside the attic, as seen in Figure 9-7, reveals the exposed backs of the modules. In order to keep the module operating temperature as low as possible, no insulation is put behind them; rather, insulation is placed in the ceiling of the house. Temperatures in the attic are high during the daytime and low at night because the modules have poor insulating value (glass and metal are excellent at transmitting heat, not at insulating). Temperature-controlled fans replace the attic's hot daytime air with cooled air from outside or elsewhere. Passive solar heating designs can be used in conjunction with solar-generated electricity by properly circulating the attic's hot air through heat exchangers or across thermal storage materials. The installation costs for an integral mount are the lowest of the four mounting schemes presented, the hardware and labor cost are comparatively low, and a substantial cost savings is achieved by replacing the entire roof material in new construction.

ELECTRICAL INSTALLATION

The electrical installation of a photovoltaic array involves a variety of tasks to ensure that various components are well interconnected and protected. The right type of electrical wire for the particular application must be selected and routed in a safe way. Connections between the wire and various terminals on a solar-cell module or system must be made securely. Proper disconnect or switching devices need to be chosen. Protection against excessive current flow requires selecting the right type of fuse or circuit breaker. The direct-current electricity produced by a photovoltaic array presents certain dangers that are less severe with conventional alternating current.

Electrical Wire

Electrical wire is composed of a low-resistance current-conducting metal surrounded by an insulating high-resistance envelope. The conducting metal typically is either copper or aluminum; gold is a superior conductor but is too expensive for most uses. Aluminum is not recommended for outdoor PV wiring because of corrosion problems. Wire is either stranded or solid, depending on the application. As shown in Figure 9-8, the cross section of a solid wire is continuous metal, while that of a stranded wire is a bundle of many smaller solid wires. Solid wire is more compact and thus has a lower resistance per given cross section and length, while stranded wire is much easier to bend and route around corners and through circuitous paths. Wire insulating material is composed of such nonconducting substances as plastic, rubber, impregnated cloth, and asbestos. Insulation is rated according to its ability to withstand high voltages, large temperature fluctuations, moisture, mildew, and chemicals such as acid, alkali, and petroleum.

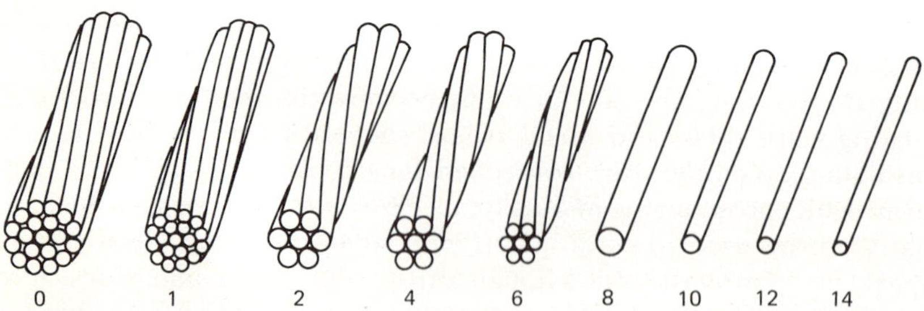

Figure 9-8 **Actual size stranded and solid wire without insulation labeled with the corresponding American wire gauge number. (From: Clyde Herrick, *Electrical Wiring: Principles and Practices,* © 1975, reprinted by permission of Prentice-Hall, Inc., Englewood Cliffs, N.J.)**

Interconnecting power wire, or *hookup wire,* has two important characteristics that need to be defined when designing and installing any electric circuit: the wire resistance or gauge and the wire insulation. Selecting the proper wire gauge and insulation for a particular application requires knowing the maximum continuous current that is expected to flow in the wire, the maximum voltage potential that will exist across the insulation, and the physical environment in which the wire will be placed.

Wire Resistance and Gauge

A conductor's resistance decreases proportionally as its cross-sectional diameter increases. The size and resistance of a wire needed for a particular application depends on the expected current that will flow in the line. The current through the conductor multiplied by the wire's resistance for the length in use indicates the voltage drop across the wire. Power loss in the line is equal to the voltage squared divided by the resistance ($P = V^2/R$); a small increase in voltage drop across the hookup wire between a power supply and its load results in a more dramatic increase in the power lost and heat generated. The maximum voltage drop generally allowed across interconnecting wire is 2% to 3% of the input voltage; any percentage in excess of this value indicates that a lower wire resistance is needed.

Wire sizes are standardized according to a scale called the American wire gauge (AWG) number system. The AWG wire table assigns each wire size a number, ranging from 40 to 0000 as shown in Table 9-1, that corresponds to a particular wire diameter and resistance. A decrease of one gauge number represents about a 25% increase in cross-sectional area and a 25% decrease in resistance.

Wire Insulation

Wire insulation is rated according to its ability to withstand such adverse conditions as high voltages, wide temperature fluctuations, moisture, chemicals, and sunlight. Insulation traditionally has been made of natural rubber; more durable plastic materials such as thermoplastic and cross-linked synthetic polymers have recently been developed that require less thickness to insulate a given voltage and are less prone to degrading in harsh environments. The National Electrical Code®* classification for different types of insulation with their respective applica-

*National Electrical Code is a registered trademark of the National Fire Protection Association, Quincy, Mass.

Table 9-1 The American Wire Gauge Table for Solid Annealed Copper Wire

	Diameter		Cross section			Resistance at 20°C*		Weight	
AWG	mm	mils	mm²	Circular mils	in²	Ω/1000 ft	Ω/km	lb/1000 ft	kg/km
0000	11.684	460.0	107.23	211600	0.1662	0.04901	0.1608	640.5	953.2
000	10.404	409.6	85.03	167800	0.1318	0.06180	0.2028	507.9	755.8
00	9.266	364.8	67.42	133100	0.1045	0.07793	0.2557	402.8	599.4
0	8.252	324.9	53.48	105500	0.08289	0.09827	0.3224	319.5	475.5
1	7.348	289.3	42.41	83690	0.06573	0.1239	0.4065	253.3	377.0
2	6.544	257.6	33.63	66370	0.05213	0.1563	0.5128	200.9	299.0
4	5.189	204.3	21.18	41740	0.03278	0.2485	0.8153	126.4	188.1
6	4.116	162.0	13.30	26250	0.02062	0.3951	1.296	79.46	118.2
8	3.264	128.5	8.366	16510	0.01297	0.6282	2.061	49.98	74.38
10	2.588	101.9	5.261	10380	0.008155	0.9989	3.277	31.43	46.77
12	2.053	80.81	3.309	6530	0.005129	1.588	5.210	19.77	29.42
14	1.628	64.08	2.081	4107	0.003225	2.525	8.284	12.43	18.50
16	1.291	50.82	1.309	2583	0.002028	4.016	13.18	7.818	11.63
18	1.024	40.30	0.8231	1624	0.001276	6.385	20.95	4.917	7.317
20	0.8118	31.96	0.5176	1022	0.0008023	10.15	33.30	3.092	4.061
22	0.6439	25.35	0.3255	642.4	0.0005046	16.14	53.81	1.945	2.894
24	0.5105	20.10	0.2047	404.0	0.0003173	25.67	84.22	1.223	1.820
26	0.4049	15.94	0.1288	254.1	0.0001996	40.81	133.9	0.7692	1.145
28	0.3211	12.64	0.08097	159.8	0.0001255	64.90	212.9	0.4837	0.7198
30	0.2548	10.03	0.05093	100.5	0.00007894	103.2	338.6	0.3042	0.4527

Source: Reference Data for Radio Engineers, I.T.&T., 1956; published with permission of the International Telephone and Telegraph Corporation.

tions and maximum operating temperatures are listed in Table 9-2. The letter code indicates the conditions various types of insulation can tolerate. The letters R, T, and X represent the insulation materials of rubber, thermoplastic, and cross-linked synthetic polymer. The number of times the letter H is found in the Code indicates how heat-resistant the material is. The letter W means that the insulation is water-resistant, and the letter N indicates that the insulation is covered with an extruded jacket of nylon.

The NEC insulation ratings are not always written on the wire or available in the literature about the conductor. If no NEC letter code is noted, it will be necessary to rely on the specific information about the wire provided by the manufacturer. Maximum voltage, temperature, and environmental ratings are usually written on the wire or listed in a catalog. If the wire has been tested and approved by a nationally known laboratory such as the Underwriters' Laboratories (UL), extra credence is added to the ratings quoted.

Ampacity

A conductor's operating temperature must be within a safe range. The heat produced in the wire as a result of internal power dissipation and the heat in the environment in which the wire is located combine to determine a conductor's operating temperature. Power loss in the line equals the current through the conductor squared multiplied by the conductor's resistance ($P = I^2R$); a doubling of the current results in a quadrupling of the power and heat dissipated. Wires that are routed in hot environments, such as in the open sun, in wire bundles, near hot water pipes, or behind solar-cell modules, will have their temperature raised by external heat. If the conductor's operating temperature becomes excessive, the wire's insulation may melt, and the exposed bare wire can pose a dangerous fire and shock hazard.

A wire's ability to withstand internal and external heat is rated according to its maximum current-carrying capacity (ampacity) and the heat tolerance of the insulating material surrounding the conductor. As the diameter of a metal conductor increases, and therefore its resistance and gauge number decreases, the maximum allowable current through the conductor naturally increases. Certain insulating materials are able to tolerate high temperatures, while other, less expensive materials are intended strictly for relatively cooler operation.

The maximum operating temperatures for different types of insulation are listed in Table 9-2, while the recommended maximum current for different gauges and insulation types is listed in Table 9-3. These National Electrical Code ratings are for wire that will carry a significant

Table 9-2 The National Electrical Code Letter System for Wire Insulation Identification

Trade name	*Type letter*	*Max. operating temp.*	*Application provisions*	*Insulation*
Heat-resistant rubber	RH	75°C 167°F	Dry location	Heat-resistant rubber
Heat-resistant rubber	RHH	90°C 194°F	Dry locations	
Moisture- and heat-resistant rubber	RHW	75°C 167°F	Dry and wet locations for over 2000 V insulation shall be ozone-resistant	Moisture and heat-resistant rubber
Heat-resistant latex rubber	RUH	75°C 167°F	Dry location	90% unmilled grainless rubber
Moisture-resistant latex rubber	RUW	60°C 140°F	Dry and wet locations	90% unmilled grainless rubber
Thermoplastic	T	60°C 140°F	Dry locations	Flame-retardant thermo-plastic compound
Moisture-resistant thermoplastic	TW	60°C 140°F	Dry and wet locations	Flame-retardant, moisture-resistant thermo-plastic

Heat-resistant thermoplastic	THHN	90°C 194°F	Dry locations	Flame-retardant, heat-resistant thermo-plastic
Moisture- and heat-resistant thermoplastic	THW	75°C 167°F	Dry and wet locations	
		90°C 194°F	Special applications within electric discharge lighting equipment. Limited to 1000 open-circuit volts or less.	Flame-retardant, moisture- and heat-resistant thermoplastic
Moisture- and heat-resistant thermoplastic	THWN	75°C 167°F	Dry and wet locations	Flame- retardant, moisture- and heat-resistant thermo-plastic
Moisture- and heat-resistant cross-linked synthetic polymer	XHHW	90°C 194°F	Dry locations	Flame- retardant cross-linked synthetic polymer
		75°C 167°F	Wet locations	

Source: Reprinted with permission from NFPA 70-1981, National Electrical Code, 1981, vol. 6, pp.70-131–70-133; this reprinted material is not the complete and official position of the NFPA on the reference subject, which is represented only by the current standard in its entirety.

Table 9-3 Allowable Ampacities of Insulated Conductors Rated 0-2000 Volts, 60° to 90°C*

Size	Temperature rating of conductor								Size
	60°C (140°F)	*75°C (167°F)*	*85°C (185°F)*	*90°C (194°F)*	*60°C (140°F)*	*75°C (167°F)*	*85°C (185°F)*	*90°C (194°F)*	
AWG MCM	Types RUW†, T†, TW†, UF†	Types FEPW†, RH†, RHW†, RUH†, THW†, THWN†, XHHW†, USE†, ZW†	Types V, MI	Types TA, TBS, SA, AVB, SIS, FEP†, FEPB†, RHH† THHN†, XHHW†,‡	Types RUW†, T†, TW†, UF†,	Types RH†, RHW†, RUH†, THW† THWN†, XHHW†, USE†	Types V, MI	Types TA, TBS, SA, AVB, SIS, RHH†, THHN†, XHHW†,‡	AWG MCM
	COPPER				ALUMINUM OR COPPER-CLAD ALUMINUM				
18	....	....	....	14	....	....	....	....	....
16	....	....	18	18	....	....	....	....	....
14	20†	20†	25	25†	....	....	....	....	....
12	25†	25†	30	30†	20†	20†	25	25†	12
10	30†	35†	40	40†	25†	30†	30	35†	10
8	40	50	55	55	30	40	40	45	8
6	55	65	70	75	40	50	55	60	6
4	70	85	95	95	55	65	75	75	4

3	85	100	110	110	65	75	85	85	3
2	95	115	125	130	75	90	100	100	2
1	110	130	145	150	85	100	110	115	1
0	125	150	165	170	100	120	130	135	0
00	145	175	190	195	115	135	145	150	00
000	165	200	215	225	130	155	170	175	000
0000	195	230	250	260	150	180	195	205	0000

CORRECTION FACTORS

Ambient temp, °C	***For ambient temperatures over 30°C, multiply the ampacities shown above by the appropriate correction factor to determine the maximum allowable load current.***								**Ambient Temp., °F**
31–40	0.82	0.88	0.90	0.91	0.82	0.88	0.90	0.91	86–104
41–45	0.71	0.82	0.85	0.87	0.71	0.82	0.85	0.87	105–113
46–50	0.58	0.75	0.80	0.82	0.58	0.75	0.80	0.82	114–122
51–60	....	0.58	0.67	0.71	....	0.58	0.67	0.71	123–141
61–70	....	0.35	0.52	0.58	....	0.35	0.52	0.58	142–158
71–80	....	....	0.30	0.41	....	....	0.30	0.41	159–176

Source: Reprinted with permission from NFPA 70-1981, National Electrical Code, 1981, vol. 6, p. 70-139; this reprinted material is not the complete and official position of the NFPA on the reference subject, which is represented only by the current standard in its entirety.

*Not more than three conductors in raceway or cable or earth (directly buried), based on ambient temperature of 30°C (86°F).

†The load current rating and the overcurrent protection for conductor types marked with an obelisk (†) shall not exceed 15 amperes for 14 AWG, 20 amperes for 12 AWG, and 30 amperes for 10 AWG copper; or 15 amperes for 12 AWG and 25 amperes for 10 AWG aluminum and copper-clad aluminum.

‡For dry locations only. See 75°C column for wet locations.

amount of electric power and current, as found in residential, commercial, and industrial wiring. Any power connection should use at least a no. 14 gauge wire, even if the current flow is considerably less than 15 A.

Wire Selection Procedure

Choosing the proper wire for a job is best done in a step-by-step way that begins with the more important criteria. First, we must assess the kind of environment the wire will be located in. Is the wire going to be routed in a wet, moist, or dry place; what kind of external temperature will it experience; will it come in contact with chemicals that might disintegrate the insulation; will the wire ever be exposed to sunlight? Based on this information, an insulation type is chosen from Table 9-2. Characteristics not covered in this table must be checked according to the manufacturer's specifications. The next design information needed is the maximum current expected to flow in the wire. The current flow between individual modules will be less than the current output from parallel branches of modules; consequently, different wire sizes will be needed for different parts of a photovoltaic array. Wires should be sized to handle at least 1 1/2 to 2 times the peak short-circuit current to account for instances of natural sunlight concentration. Once the insulation type needed and the expected current flow are known, the wire gauge number can be looked up in Table 9-3. As a final check, calculate the voltage drop across the wire. An estimate of the wire length needed for the particular application will have to be made. Multiplying the intended wire length by the resistance per unit length for the gauge chosen (found in Table 9-1) will give us the total wire resistance. The expected current multiplied by the wire resistance equals the wire voltage drop. Is the wire voltage drop greater than 2% or 3% of the source voltage that will be attached to the wire? If yes, the next smaller gauge wire should be considered.

Let us go through a design example and choose the proper wire for interconnecting modules of a roof-mounted solar-cell array that has an operating voltage of 240 V and a peak short-circuit current of 10 A. Interconnecting wire between externally mounted solar-cell modules will be exposed to the elements, and therefore the insulation must be able to tolerate a wet environment. The temperature behind a solar-cell module will fluctuate quite dramatically, so an approximate temperature range of between −30°C and +100°C is needed. The wire generally is shaded, but it might come in contact with the various chemicals found in roofing material or the earth and therefore must be resistant to them. Given these criteria, we will choose a XHHW type insulation as found in Table 9-2. The need for an insulation that is resistant to chemicals will prompt us to check wire manufacturers' specifications that have been certified by a recognized laboratory (underground feeder cable that has been tested by the Underwriters' Laboratories will meet our needs). We know that the hookup wire may have as much as 15 A flowing in it (after allowing for sunlight concentration extremes). Looking in Table 9-3, we find that the AWG number for maximum 15-A current flow and a XHHW insulation rating is 14. We expect to use about 100 ft of no. 14 gauge wire. Looking up the resistance for no. 14 gauge wire in Table 9-1, we find it to be 2.53 Ω/1000 ft. Our 100-ft wire's resistance will thus be 2.53 Ω/1000 ft multiplied by 100 ft, or 0.253 Ω. If 15 A of current flows through 0.253 Ω of resistance, we will have a 3.80-V voltage drop across the line. The operating voltage of our array is 240 V; a 3.80-V voltage drop represents only 1.6% of the total, which is well within the 2% to 3% limit. Number 14 gauge wire with an XHHW rating that is specified to withstand the types of chemicals found in the immediate environment is acceptable for this particular use.

Wire Routing

The routing of wire in a solar-cell system must observe various protection and safety considerations. Good wiring practice require that conductors be sufficiently protected from the environment, that they be routed with strain relief and without sharp bends, and that they be neatly arranged and labeled.

Electrical wire often is sealed or routed inside an additional protective cover, such as multiconductor cable, conduit, or raceway, to provide added insulation or protection. A common practice is to put two or more insulated wires inside another insulation jacket. For instance, internal residential and commercial wiring is mostly done with Romex cable that has three wires in one plastic sheath (positive, negative or neutral, and ground). Multiconductor cables are labeled with two numbers separated by a slash or dash; the first number denotes the gauge number of a single conductor, and the second number indicates how many conductors are in the cable (a cable labeled 14/3 has three AWG no. 14 wires in it). Conduit is a metal pipe, either rigid or flexible, through which one or more wires are snaked. A raceway is a trough, usually suspended from the ceiling, in which wires are laid. Wires carrying alternating current should not be routed in the same protective cover as wires carrying direct current; induced current from the AC line into the DC line can produce an undesirable ripple on the DC voltage.

Wires should be routed with sufficient strain relief or slack and without any tight bends. The normal flexing of materials or movement of parts during operation or maintenance may pull on a connected wire. If there is no slack in the wire, it or its connection may be strained to the point of breaking. Routing wire with tight bends, especially near sharp edges, is an invitation for the insulation to crack and the exposed circuit to present a hazard. A good practice is to place a strain relief loop of wire near each connection and to corner wire with a reasonably large arc.

Troubleshooting and repair will often require retracing a wire in search of a fault. If wires are not properly organized and labeled, this will be very difficult. Electrical convention states that wires with a positive voltage on them should have black insulation, while negative or grounded neutral wires should be white. A ground wire normally does not carry current and generally has green insulation or is bare without insulation. Labels with a number or letter code are often put on or near a wire to denote its origin or destination.

The National Electrical Code provides minimum standards for electrical wiring that are intended to promote safety from fire and shock hazards. The enforcement and interpretation of the code vary from one locality to the next. It therefore is important to check with local inspectors and government agencies before initiating any wiring plan.

Electrical Connections

Interconnecting power wires in a solar-cell array or system must be attached to terminals on a module or circuit in a secure and reliable way; however, maintenance or repair occasionally will require disconnecting these wires with the least possible complication. Terminating hookup wire with various types of lugs or connectors will satisfy both of these considerations: they make a tight metallic contact between two portions of a circuit, and they allow for simple disconnection.

Terminal Lugs

Lugs can be either U- or O-shaped metal pads that are crimped onto the bare metal end of a wire or clamps that are attached to the bare wire by a screw-down vise, as shown in Figure 9-9. In either case, the lug is attached to the desired circuit terminal by fastening it under a screw. To secure crimp lugs to a wire, the insulation is stripped back sufficiently to fit the lug over the bare wire, then the back end of the lug is crushed around the wire using a special crimping tool. Clamping lugs are generally used for lower-gauge wires (about AWG no. 6 or less) and are attached to the bare end of a wire by squeezing it under a screw-down vise.

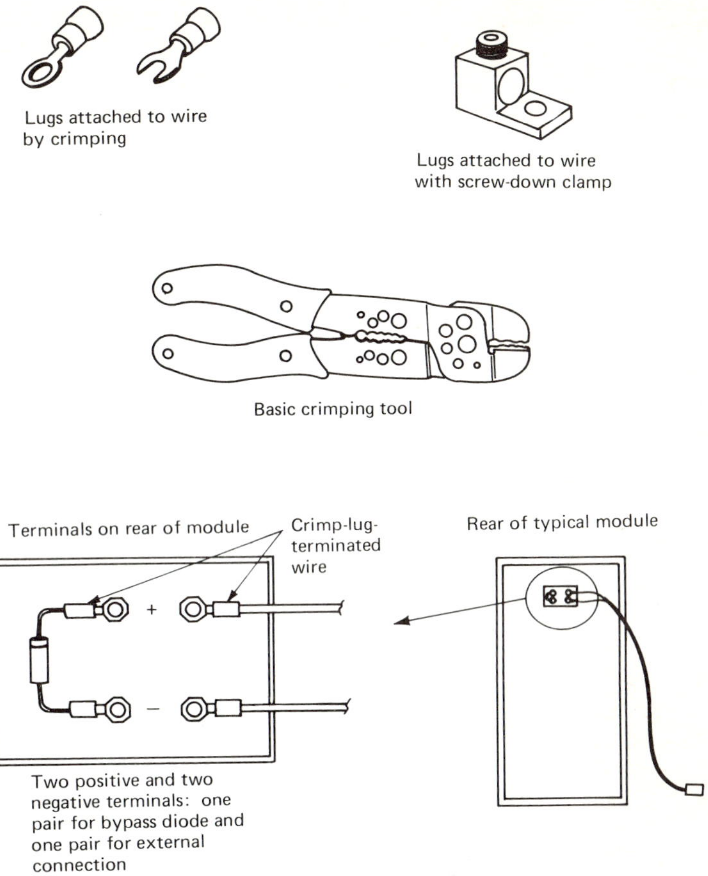

Figure 9-9 Various types of wire terminating lugs and their typical applications for wiring a photovoltaic module.

Connectors

Connectors terminate a wire with either male metal prongs or female metal receptacles that easily mate by being pushed together. Photovoltaic connectors should have a unique design that prevents them from being interchanged with other common household plugs. Since voltage may be present on both sides of a connection, neither the male nor the female connector should have "live" metal parts exposed that could present a shock hazard, such as protruding prongs. The contacts of a photovoltaic array connector should be set apart so that the rated current can easily be interrupted without excessive arcing when two mated connectors are pulled apart. A connector used between solar-cell modules should have

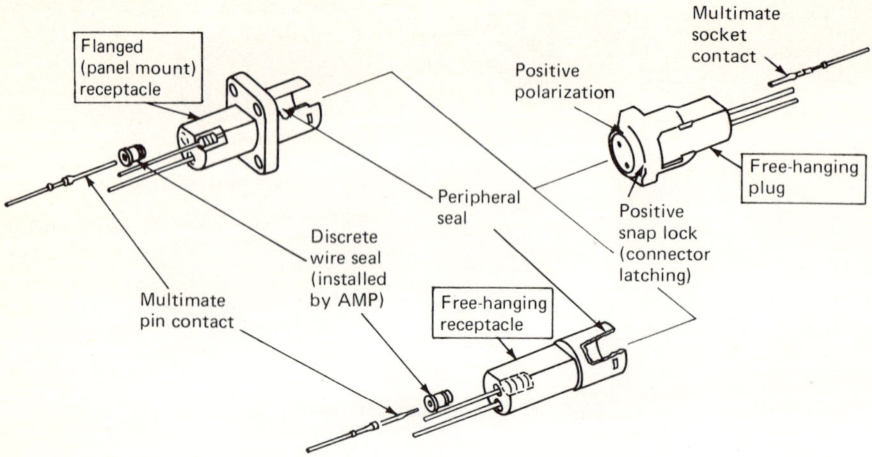

(a)

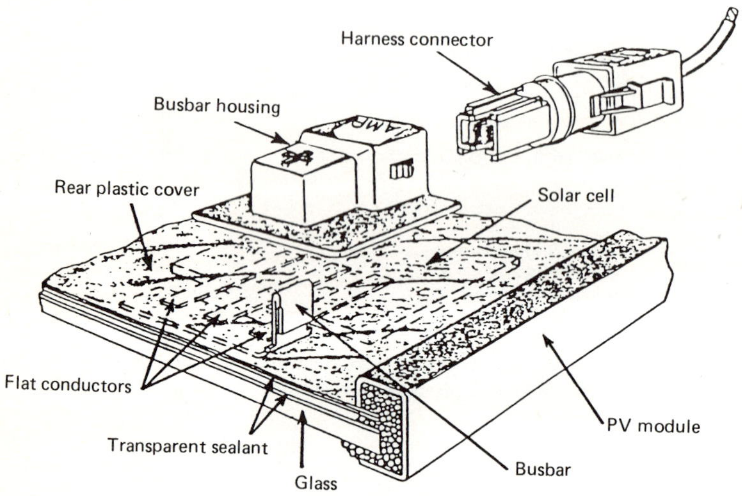

(b)

Figure 9-10 **Connectors suitable for use in photovoltaic-array wiring. (*a*) AMP Econosea® Connector (© AMP Inc., Harrisburg, Pa.; reprinted with permission). (*b*) AMP SOLARLOK® Connector (© AMP Inc., Harrisburg, Pa.; reprinted with permission.)**

a locking mechanism with a weather seal to prevent the connection from falling apart and moisture corroding the metal contacts; locking is usually achieved with a snap or twist-on mechanism, while weather sealing is typically accomplished with a rubber O ring. Connectors generally are attached to the bare end of a wire by being crimped on with a special tool. Examples of connectors that are suitable for photovoltaic-array DC interconnections are shown in Figure 9-10.

Electrical connections should be made with similar metals, copper with copper and aluminum with aluminum. The atomic structures of dissimilar metals interact in a way that increases corrosion when the metals are placed in close contact.

Circuit Protection and Interruption Devices

Circuit protection and interruption, whether for the purpose of overcurrent prevention or for switching, require that the circuit be broken at some point. Breaking a circuit presents an especially acute problem when direct current is flowing, as electric arcing across a sudden gap created in a circuit presents various fire and shock hazards that must be prevented.

Arcing occurs when the insulation between two electrical points of different potential breaks down; the heat produced when current flows across a high-resistance breakdown path can easily start a fire if any flammable materials are in the vicinity. Once an arc begins, it often is sustained because the arcing itself can reduce the resistance between two points. For instance, if the voltage between two electrodes builds up to a level that forces current to arc through air from one conductor to the next, the heat and ionization of air molecules produced by the arc reduce the air's resistance. A decrease in the resistance of the arcing path causes more current to flow until it stabilizes in a continuous state. Arcing is interrupted only if the current flow in one direction is stopped.

Direct current is far more prone to arcing across a gap in a conductor than is alternating current. Alternating current by its nature has many interruptions in current flow as it switches back and forth. The utility 60-cycle-per-second alternating current momentarily has zero current flowing 120 times a second as it switches direction. These momentary interruptions make it more difficult for an arc to sustain itself, and therefore AC arcing only occurs under conditions of high voltages or small gaps. Direct current, on the other hand, flows continuously in one direction and will sustain an arc at lower voltages and across larger gaps.

The higher incidence of arcing in direct-current circuits leads to a requirement that any device intended to interrupt a DC circuit, such as a switch or a fuse, must be specifically designed to inhibit arcing. Most electrical devices are designed for AC applications. AC voltage and current ratings on switches and fuses, however, are inadequate for DC circuits and consequently must not be trusted. All circuit protection and interruption devices used in DC circuits must have a DC voltage and current rating that exceeds the electricity expected in the circuit.

Switching

The two basic switching devices for interrupting a circuit are the *switch* and the *relay*. The switch opens or closes one or more sets of contacts with a lever action. The relay controls one or more sets of contacts by energizing or de-energizing a magnetic field. DC switching with a switch or a relay requires that the contacts open wide with a quick and decisive motion to prevent an arc from occurring. Even unsustained arcs are capable of gradually pitting and destroying the contacts and therefore must be prevented if at all possible.

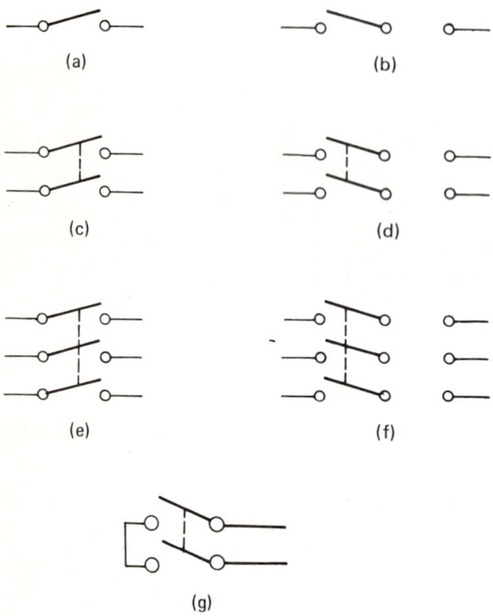

Figure 9-11 Common switch pole and throw arrangements: (*a*) single-pole single-throw (SPST), (*b*) single-pole double-throw (SPDT), (*c*) double-pole single-throw (DPST), (*d*) double-pole double-throw (DPDT), (*e*) triple-pole single-throw, (*f*) triple-pole double-throw, (*g*) arc-suppressing DC-switch wiring arrangement.

The number and arrangement of the contacts on a switch or a relay is designated by the number of *poles* and *throws* it has. The number of poles indicates how many pairs of contacts are opened or closed simultaneously, while the number of throws designates how many contact positions the switch is capable of resting in. Typical pole and throw arrangements are shown in Figure 9-11. A wiring scheme for reducing DC arcing arranges a double-pole double-throw switch in a configuration as shown in Figure 9-11; the circuit is opened or closed in two or more places simultaneously.

The Switch The DC-rated switch makes or breaks a contact quickly with a spring-loaded mechanism that snaps the contacts together or apart. The voltage rating indicates the maximum voltage that the open contacts will sustain without arcing. The current rating is an indication of how large the contacts are, how low the resistance is, and therefore how much current they can sustain without overheating. A typical ex-

Figure 9-12 **External disconnect switch with fuse protection used to disconnect a 6-kWp PV array from internal house wiring. (Photo by author.)**

ternal disconnect switch for a PV array is shown in Figure 9-12. When a switch is used with a resistive load, the current rating represents the maximum current allowed through the switch. However, when a switch is used with an inductive load, such as a motor, the current rating is twice the allowable current. As discussed in Chapter 7, when a voltage initially is put across an inductive load, for a brief instant a large transient current flows that is considerably larger than the normal operating current.

The Relay The mechanical relay opens and closes a set of contacts by controlling their position with a magnetic field. The relay's basic components are a coil of wire, one or more pair of contacts, and a spring that holds the contacts apart (or sometimes together, whichever the case might be). When current flows through the coil, it produces an electromagnet that pulls the contacts together (or apart) by overcoming the spring's tension. The contacts can be either two metal pads or a metal electrode and a pool of mercury. A pair of contacts are said to be *normally open* (NO) or *normally closed* (NC) depending on where they rest when the coil is de-energized. The same precautions must be taken with relay contacts as with a switch.

A relay, in effect, involves two switches; one switch is needed to turn the coil current off and on, while the other switch is controlled by the coil's magnetic field. The advantage of a two-switch device, such as a relay, over one switch is apparent when dealing with large amounts of power. A relay makes it possible to switch high-power circuits with a low-power switch. The relay contacts can be rated to handle large amounts of current and voltage, while the contacts in series with the coil need operate only with small quantities of electricity. A relay isolates the dangerous levels of electric power from the controls that are operated by expensive electronics or vulnerable people.

Overcurrent Protection

Photovoltaic arrays need overcurrent protection only when there is a danger that utility or storage power might feed back into the array. Unlike the electric utility, photovoltaic cells are naturally current-limited. The utility is a voltage source (i.e., it provides all the current needed to maintain a fixed voltage), while a photovoltaic array is a fairly good current source (it tries to adjust the voltage level to maintain a particular current). A short circuit in the array will not yield more than the maximum short-circuit current; a short across a battery or the utility will cause the current to approach infinity.

The parameter that can rise to dangerously high levels with a voltage source is the current, while with a current source it is the voltage. Protection in a voltage-source circuit involves preventing the current from climbing too high; hence the utility wiring is filled with fuses and current-sensing circuit breakers. In a current-source circuit, the voltage must be restricted. However, a photovoltaic cell is a poorer current source than the utility is a voltage source. A solar cell's voltage drops off quickly near open-circuit voltage, as is evident in the *I/V* curves shown in Chapter 4. If all the wire and components connected to a solar-cell array are rated for the array's open-circuit voltage, no voltage protection is needed.

There are two basic circuit protection devices that detect and interrupt excessive current flow in a circuit. A *fuse* is an intentional weak link that self-destructs and opens up when the current flow through it exceeds its rating. A *circuit breaker* similarly opens the circuit when the current

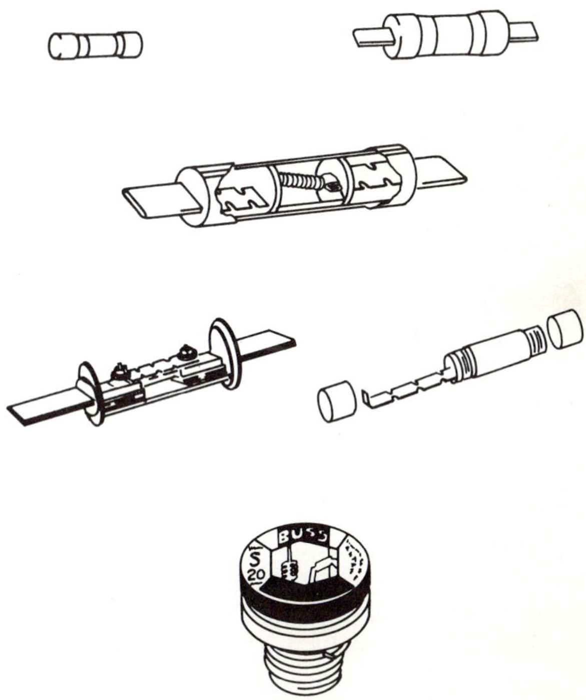

Figure 9-13 **Various types of fuses, showing some packaging designs. (From: Clyde Herrick, *Electrical Wiring: Principles and Practices*, © 1975, reprinted by permission of Prentice-Hall, Inc., Englewood Cliffs, N.J.)**

is too high, but it does so by tripping an internal switch that later can be reset. A circuit breaker is more expensive than a single fuse, but does not need to be replaced after each failure. However, malfunctions in overcurrent protection are more likely to occur with a circuit breaker than with a fuse (as is evident in their respective designs).

Fuses A fuse is composed of two electrodes with a metal strip between them, as shown in Figure 9-13. The metal strip maintains a short circuit between the two electrodes at continuous current levels beneath its rating. If the current flow exceeds the prescribed rating of the fuse for at least a few seconds and sometimes less, the metallic strip melts and opens up the circuit. If the electrodes are too close together, arcing may occur after the fuse has blown, negating its purpose. Fuses therefore have maximum voltage ratings that indicate the highest voltage a blown fuse can sustain without arcing.

Circuit Breakers A circuit breaker is composed of a switch and an automatic trip mechanism that opens the switch when excessive current is flowing through the circuit. The toggle switch on a circuit breaker usually has three positions, ON, TRIPPED, and OFF; a breaker that has tripped as a result of overcurrent is usually reset to the ON position by turning the toggle switch first OFF and then ON. The switching contacts in a circuit breaker require the same precautions observed with a conventional switch.

The automatic trip mechanism is activated either by heat or by a magnetic field. A thermal circuit breaker opens a circuit by using a bimetallic strip that bends as the strip's current-induced temperature increases, while a magnetic circuit breaker makes use of a coil's current-induced magnetic field. Either the bending of a bimetallic strip or the magnetic attraction of a coil causes a latch to be raised and a spring-loaded arm to open a set of contacts. A thermal relay responds to excessive current more slowly than a magnetic relay because it takes time for the bimetallic strip to heat up and bend.

The Diode Selecting a diode for a particular application requires defining two parameters that describe how well the diode performs its function of permitting current flow in one direction and preventing it in the other. The maximum continuous forward current and the peak inverse voltage that a diode will experience in a circuit must be known. The maximum forward current rating indicates how many amperes can safely flow through the diode in the forward direction. The peak inverse voltage rating refers to the maximum reverse-polarity voltage a diode can tolerate without breaking down and allowing significant amounts of current to flow.

Diodes play two important functions in photovoltaic-array circuits. Shunt or bypass diodes are placed across each solar-cell module to divert current in the event that an open circuit occurs somewhere in the module. Blocking diodes are placed at the end of a string of modules to prevent external power from forcing its way back into the string; another string of modules or an electric storage system wired in parallel can inject current into a malfunctioning or unilluminated solar-cell string if reverse current is not blocked.

The blocking diode will have current flowing in it on a daily basis that could reach a maximum level around the string's peak short-circuit current (at one sun or a clear noon insolation). The shunt diode across a module similarly will have to handle the string's maximum current in the event that a module should fail or is shaded. Both the blocking diode and the shunt diode should be rated to handle about 1-1/2 to 2 times the string's peak short-circuit current to handle instances of natural sunlight concentration. The peak inverse voltage rating (PIV) of a blocking diode must well exceed the string's open-circuit voltage, while a shunt diode's voltage rating must only be considerably higher than a module's open-circuit voltage.

Diodes generally come in one of two different packages: a cylindrical case with axial leads or a stud mount case with an extruding bolt for mounting on a metal heat sink, as shown in Figure 8-17. Diodes designed for high-power applications (high forward current ratings) usually come in a stud mount case so that a heat-sink mount can spread the diode's heat out over a large area and thus allow for more rapid cooling. Proper procedures must be followed to heatsink these diodes and electrically

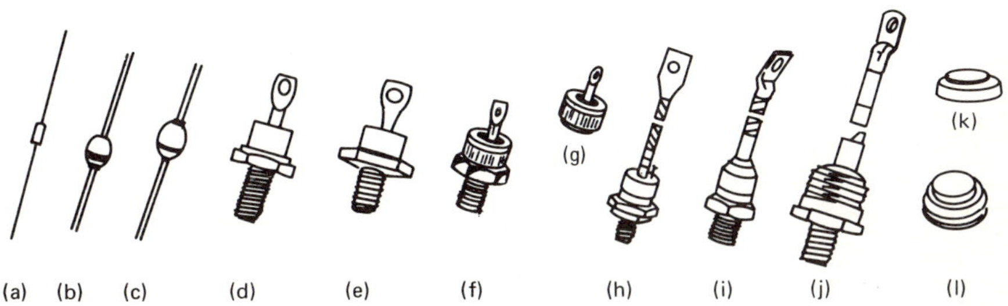

Figure 9-14 Common silicon diode packages of 0.25 to 150A, 50 to 1500 V. (© General Electric Company; reprinted with permission.)

isolate their electrodes to prevent shock hazards. Lower-power axial-lead diodes are usually terminated with crimp lugs and are mounted under a screw on a terminal. Examples of axial-lead diodes are shown in Figure 9-14*a* to *c*, while stud-mount packaged diodes are shown in Figure 9-14*d* to *j*.

Bibliography

This book relied on M.I.T. Lincoln Laboratory reports and conference papers too numerous to list here. To obtain a list and copy of these documents, contact:

The National Technical Information Service
U.S. Department of Commerce
Springfield, Va. 22161

Power Systems Group, Ametek, Inc.: *Solar Energy Handbook, Theory and Application,* Chilton, Radnor, Pa., 1979.

Backus, Charles E.: *Solar Cells,* IEEE Press, New York, 1976.

Basile, Paul S.: *Energy Demand Studies: Major Consuming Countries,* M.I.T. Press, Cambridge, Mass., 1976.

Bechtel National, Inc.: *Handbook for Battery Energy Storage in Photovoltaic Power Systems, Final Report,* Sandia Laboratory, Albuquerque, N. Mex., 1979.

Cox, Charles H.: "Power-Producing Homes: Making the Utility Connection," *Solar Age,* December 1979.

Dickinson, William C., and Paul Cheremisinoff: *Solar Energy Technology Handbook,* Marcel Dekker Inc., New York, 1980.

Doule, John M.: *An Introduction to Electrical Wiring,* Reston Publishing Company, Prentice-Hall, Reston, Va., 1975.

Duffie, John A., and William Beckman: *Solar Engineering of Thermal Processes,* Wiley, New York, 1980.

Forman, S. E.: *Detection of Electrically Failed Photovoltaic Modules at M.I.T. Lincoln Lab Test Sites,* Massachusetts Institute of Technology Lincoln Laboratory, Lexington, Mass., 1981.

———: *Summary of Field Experience of Photovoltaic Modules at Various M.I.T. Lincoln Lab Test Sites,* Massachusetts Institute of Technology Lincoln Laboratory, Lexington, Mass., 1978.

———: *Visual Defects in Terrestrial Photovoltaic Modules,* Massachusetts Institute of Technology Lincoln Laboratory, Lexington, Mass., 1978.

Goff, H. C., T. S. Chan, D. W. Allred, and J. A. Nichols: *Identifying Electrical Loads Which Can Use Direct Photovoltaic DC Power,* Sandia Laboratory, Albuquerque, N. Mex., 1981.

Gupta, Y. P., and S. K. Young: *Design Handbook for Photovoltaic Power Systems,* Sandia Laboratory, Albuquerque, N. Mex., 1981.

Herrick, Clyde N.: *Electrical Wiring: Principles and Practices,* Prentice-Hall, Englewood Cliffs, N.J., 1975.

Hoffman, A. R., and C. R. Maag: *Photovoltaic Module Soiling Studies May 1978–October 1980,* California Institute of Technology Jet Propulsion Laboratory, Pasadena, Calif., 1980.

Howell, Yvonne, and Justin A. Bereny: *Engineer's Guide to Solar Energy,* Solar Energy Information service, San Mateo, Calif., 1979.

Jet Propulsion Laboratory: *Workshop on Flat Plate Photovoltaic Module and Array Circuit Design Optimization,* California Institute of Technology, Pasadena, Calif., 1980.

Johnson, W. D.: *Solar Voltaic Cells,* Marcel Dekker, Inc., New York, 1980.

Kirpich, Aron, G. O'Brien, and N. Shepard, *Electric Power Generation: Photovoltaics,* General Electric Company, Philadelphia, 1980.

Kreider, Jan F., and Frank Kreith: *Solar Energy Handbook,* McGraw-Hill, New York, 1981.

Landsman, E. E.: *Maximum Power Trackers for Photovoltaic Arrays,* M.I.T., Lexington, Mass., 1978.

———: *Static Inverters for Use in PV Power Systems,* M.I.T., Lexington, Mass., 1979.

Lyon, E. F.: *Discussion of Technical Issues Associated with the Interconnection of Residential Solar Electric Energy Sources with the Utility System,* M.I.T., Lexington, Mass., 1981.

Matlin, R. W.: *Design Optimization and Performance Characteristics of a Photovoltaic Microirrigation System for Use in Developing Countries,* M.I.T., Lexington, Mass., 1979.

Maycock, Paul D., and Edward N. Stirewalt: *Photovoltaics: Sunlight to Electricity in One Step,* Brick House Publishing Co., Andover, Mass., 1981.

Mazria, Edward: *Passive Solar Energy Book,* Rodale, Emmaus, Pa., 1979.

Mehalick, E. M., G. F. Tully, J. Johnson, J. Parker, and R. Felice, *The Design of a Photovoltaic System for a Passive Design Northeast All-Electric Residence,* Sandia Laboratory, Albuquerque, N. Mex., 1982.

———,———,———, N. Truncellito, R. Schaeffer, and J. Parker: *The Design of a Photovoltaic System for a Southeast All-Electric Residence,* Sandia Laboratory, Albuquerque, N. Mex., 1982.

———, G. O'Brien, and G. F. Tully: *The Design of a Photovoltaic System for a Southwest All-Electric Residence,* Sandia Laboratory, Albuquerque, N. Mex., 1980.

———, G. F. Tully, J. Johnson, N. Truncellito, and R. Schaeffer: *The Design of a Photovoltaic System for a Temperate Climate All-Electric Residence,* Sandia Laboratory, Albuquerque, N. Mex., 1982.

Merrigan, Joseph A.: *Sunlight to Electricity,* M.I.T. Press, Cambridge, Mass., 1975.

Monegon, Ltd.: *Solar Electricity: Making the Sun Work for You,* Gaithersburg, Md., 1981.

National Fire Prevention Association: *National Electrical Code,* Boston, 1975.

Neville, Richard C.: *Solar Energy Conversion: The Solar Cell,* Elsevier, New York, 1978.

Nichols, B. E.: *The Carlisle House: An All Solar Electric Residence,* Massachusetts Institute of Technology Lincoln Laboratory, Lexington, Mass., 1981.

———: *A Solar Photovoltaic Power System for a Radio Station,* Massachusetts Institute of Technology Lincoln Laboratory, Lexington, Mass., 1980.

Office of Technology Assessment: *Application of Solar Technology to Today's Energy Needs,* United States Congress, Washington, D.C.

Palz, Wolfgang: *Solar Electricity, An Economic Approach to Solar Energy,* Butterworths, London, 1978.

Pulfrey, David L.: *Photovoltaic Power Generation,* Van Nostrand Reinhold, New York, 1978.

Rasmussen, N. E., and H. M. Branz: *The Dependence of Delivering Energy on Power Conditioning Electrical Characteristics for Utility-Interactive Systems,* M.I.T., Lexington, Mass., 1981.

Rauschenbach, H. S.: *Solar Cell Array Design Handbook,* Van Nostrand Reinhold, New York, 1960.

Romer, Robert H.: *Energy: An Introduction to Physics,* Freeman, San Francisco, 1976.

Russell, Miles C.: "An Apprentice's Guide to Photovoltaics," *Solar Age,* July 1981, p. 32.

Smith, G.: *Storage Batteries,* Pitman, London, 1964.
Solarex Corporation Technical Staff: *Making & Using Electricity from the Sun,* Tab Books, Blue Ridge Summit, Pa., 1979.
Solman, F. J.: *Photovoltaic Systems Performance Experience,* M.I.T., Lexington, Mass., 1981.

APPENDIX A

STATE OF THE ART: PHOTOVOLTAIC SYSTEMS IN OPERATION

TABLE A-1 PV-Powered Railroad Track System Description.

	Application #1	*Application #2*
APPLICATION:	Railroad Track Circuit	Railroad Signal Lamp
LOCATION:	Lake Pontchartrain, Louisiana	Lake Pontchartrain, Louisiana
VOLTAGE:	2.8 Vdc nominal	12 Vdc nominal
LOAD:	40 Ah/day; track occupied 6 hours/day, 5 days/week	30 Ah/day; track occupied 6 hours/day, 5 days/week
SOLAR ARRAY ELECTRICAL SPECIFICATIONS (at battery charging point):	3 Vdc, 9.1 A, 27.3 W	13.8 Vdc, 7.8 A, 108 W
DIMENSIONS:	48″ × 32″ × 3″	62″ × 51″ × 3″
WEIGHT:	70 lb, including hardware	210 lb, including hardware
BATTERY:	786 Ah, 2.8 Vdc, nickel-cadmium type	554 Ah, 12 Vdc, lead-acid type
INSTALLED:	Summer 1976	Summer 1976

Source: Solar Power Corporation.

Figure A-1 PV-powered railroad track circuit. (Photo courtesy Solar Power Corporation.)

Table A-2 PV-Powered Navigation Equipment System Description.

APPLICATION:	Aids to Navigation
LOCATION:	Offshore oil platform, Gulf of Mexico
VOLTAGE:	24 Vdc Nominal
LOAD:	Horn—4 A, 10% duty cycle; 0.06 A, 90% duty cycle; 24-hour operation TOTAL LOAD: 10.9 Ah/day
SOLAR ARRAY:	26.4 Vdc, 3.2 A, 84.5 W
DIMENSIONS:	50″ × 50″ × 2″
WEIGHT:	130 lb
BATTERY:	335 Ah, 24 Vdc (12 cells); stationary lead-calcium cells
INSTALLED:	Summer 1975

Source: Solar Power Corporation.

Figure A-2 PV-powered navigation equipment. (Photo courtesy Solar Power Corporation.)

Table A-3 PV Power Microwave Repeater Station System Discription

APPLICATION:	Microwave Repeater Station
LOCATION:	Challis, Idaho
VOLTAGE:	24 Vdc Nominal
LOAD:	5.0 A Continuous TOTAL LOAD: 120 Ah/day
SOLAR ARRAY:	27.6 Vdc, 30 A, 816 W
DIMENSIONS:	Six separate arrays, each measuring 75″ × 50″ × 2″
WEIGHT:	1130 lbs
BATTERY:	2190 Ah, 24 Vdc (12 cells); stationary lead-calcium cells.

Source: Solar Power Corporation.

Figure A-3 PV-powered microwave repeater station. (Photo courtesy Solar Power Corporation.)

Table A-4 PV Power Water Pumping System Description.

APPLICATION:	Remote water Pumping
LOCATION:	Queen's Well, Papago Indian Tribe, Arizona
VOLTAGE:	120 Vdc nominal
LOAD:	120-V, 3-hp Honeywell motor pump; daytime only operation
SOLAR ARRAY ELECTRICAL SPECIFICATIONS: (at battery charging point)	195 Vdc, 13.66 A, 2664 W, @ 25°C
AREA:	26.75 m^2
BATTERY:	Not required; water storage used
INSTALLED:	November 1981

Source: Trisolar Corp., Bedford, Mass.

Figure A-4 **PV power water pumping. (Photo courtesy Trisolar Corp., Bedford, Mass.)**

Table A-5 Carlisle House System Description.

APPLICATION:	All solar-electric residential house
LOCATION:	Carlisle, Massachusetts
PV ARRAY OUTPUT:	7.3 kW at 46°C; 190 V × 4.25 A
PV ARRAY CONFIGURATION	
ELECTRICAL:	9 modules in series by 14 in parallel
PHYSICAL:	Residential roof standoff mount: 7 modules high by 18 modules wide
PV ARRAY AREA:	99 m^2
ENERGY STORAGE:	Utility grid
BACKUP POWER:	Utility; passive and active solar heating
POWER CONDITIONING:	8-kVA line-commutating inverter with fixed voltage input
LOAD:	38,500 Btu/h heat pump (at 8.5°C) Backup resistive heating Electric cooking, clothes washing and drying, hot water (as backup to 10-m^2 active solar hot water collectors), and normal residential loads
OPERATION TIME:	June 1981–present

Figure A-5.1 Carlisle House residential photovoltaic system. (Photo courtesy M.I.T. Lincoln Laboratory.)

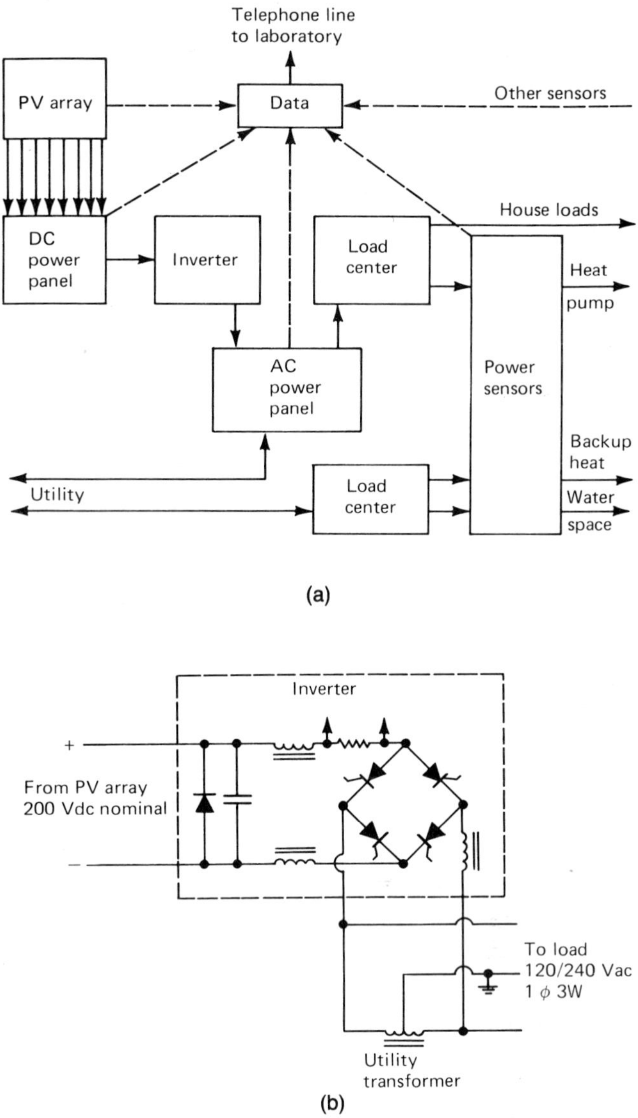

Figure A-5.2 **Carlisle House: (*a*) PV-system circuit diagram and (*b*) utility-inverter connection. (M.I.T. Lincoln Laboratory.)**

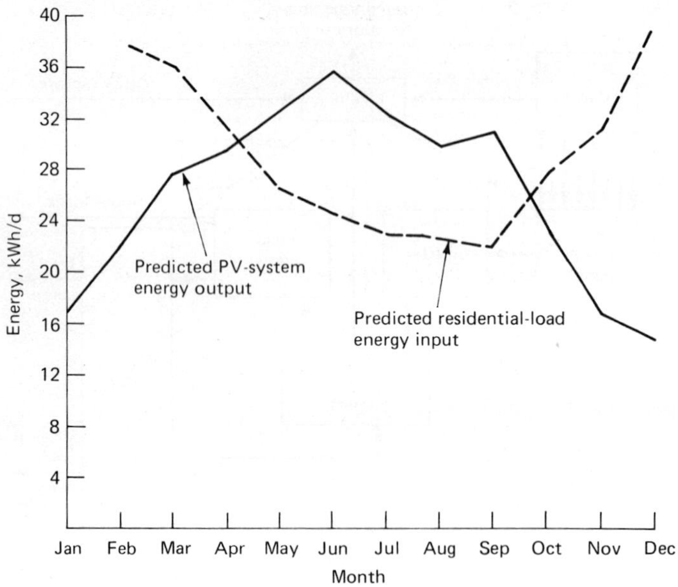

Figure A-5.3 Carlisle House predicted PV power output and load input.

Table A-6 WBNO Radio Station System Description

APPLICATION:	Solar-powered 500-W daytime AM radio station (WBNO)
LOCATION:	Bryant, Ohio
PV ARRAY OUTPUT:	15 kWp at 20°C; 130 V × 115 A (approx.)
PV ARRAY CONFIGURATION:	
ELECTRICAL:	50 branch circuits: 4 modules in series by 2 in parallel in series with 4 in series by 2 in parallel
PHYSICAL:	Rack mount; 106 1.22 m × 2.44 m panels in 14 rows of 5–8 panels each
PV ARRAY AREA:	Array: 296.3 m^2 Land: 1/4 acre
ENERGY STORAGE:	Lead-acid batteries; 40 kWh capacity 128 Vdc nominal discharge voltage
BACKUP POWER:	Utility
POWER CONDITIONING:	5-kW dc-to-ac inverter; dc-to-dc converter
LOAD:	AM transmitter, radio station lighting, control, and general electricity needs
OPERATION TIME:	September 1979 to present

Figure A-6.1 **PV-powered daytime AM radio station (WBNO). (Photo courtesy M.I.T. Lincoln Laboratory.)**

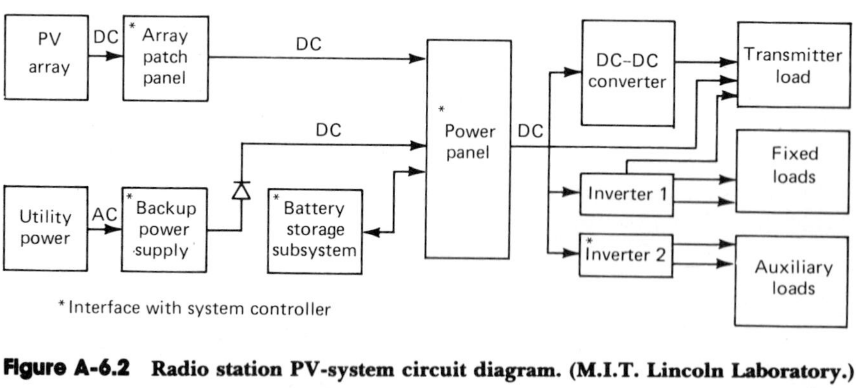

Figure A-6.2 **Radio station PV-system circuit diagram. (M.I.T. Lincoln Laboratory.)**

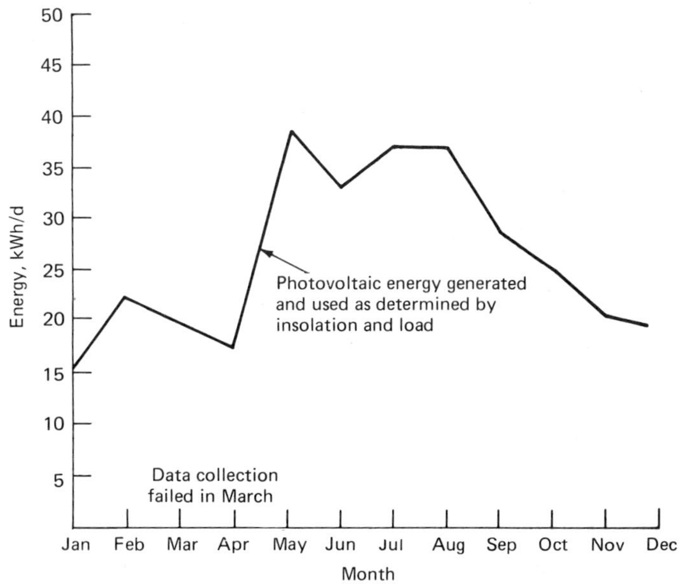

Figure A-6.3 Radio station PV-system energy output during 1980. *Note:* The size of the battery and the load forced this system to discard at least 30 percent of the available solar photovoltaic energy.

Table A-7 Agricultural Station System Description.

APPLICATION:	Solar-powered agricultural station with 80-acre cornfield irrigation, fertilizer production, and grain drying
LOCATION:	Mead, Nebraska
PV ARRAY OUTPUT:	25 kWp at 20°C; 120 V × 205 A (approx.)
PV ARRAY CONFIGURATION:	
ELECTRICAL:	42 branches: 4 modules in parallel by 9 in series 35 branches: 2 modules in parallel by 10 in series
PHYSICAL:	Rack mount: 28 2.5 m × 7.6 m panels in two rows of 14 panels each.
PV ARRAY AREA:	Array: 532 m^2 Land: 1/3 acre
ENERGY STORAGE:	Lead-acid battery, 90 kWh capacity 120 V nominal discharge voltage
BACKUP POWER:	Utility
POWER CONDITIONING:	
LOAD:	20-hp water pump 10-hp air blower 3-kW fertilizer generator 4-hp sweep auger 3-hp spread auger 10-hp load/unload auger
OPERATION TIME:	January 1978 to present

Figure A-7.1 PV-powered agricultural station. (Photo courtesy M.I.T. Lincoln Laboratory.)

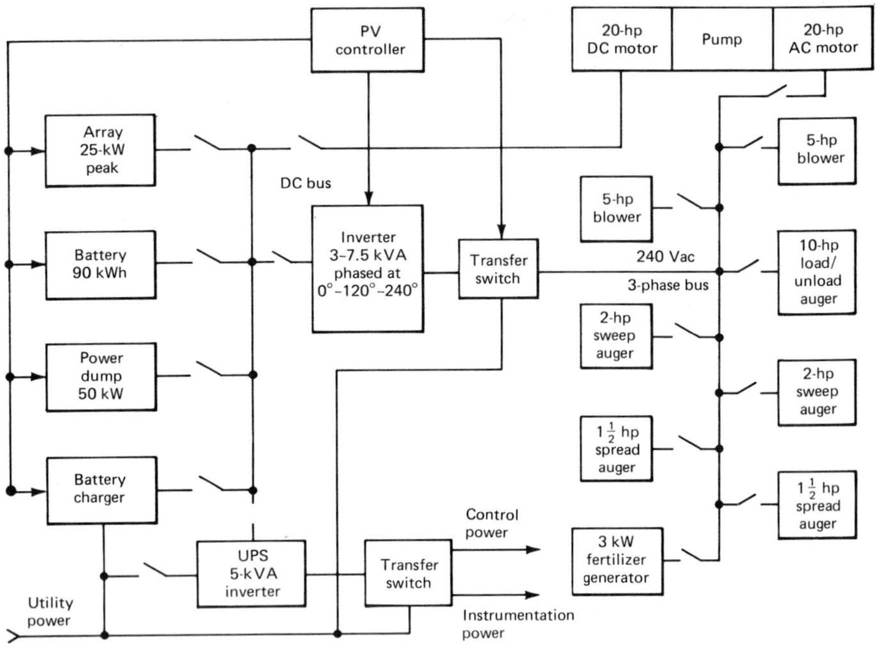

Figure A-7.2 Agricultural station PV-system circuit diagram. UPS-uninterrupted power supply. (M.I.T. Lincoln Laboratory.)

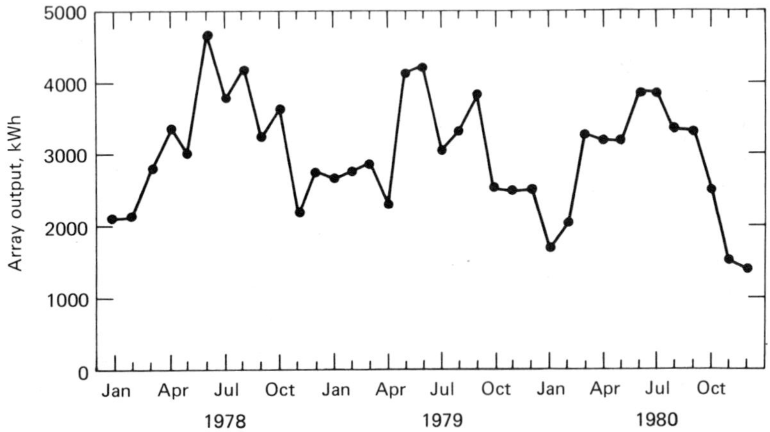

Figure A-7.3 Agricultural station PV-system energy output between 1978 and 1980. (M.I.T. Lincoln Laboratory.)

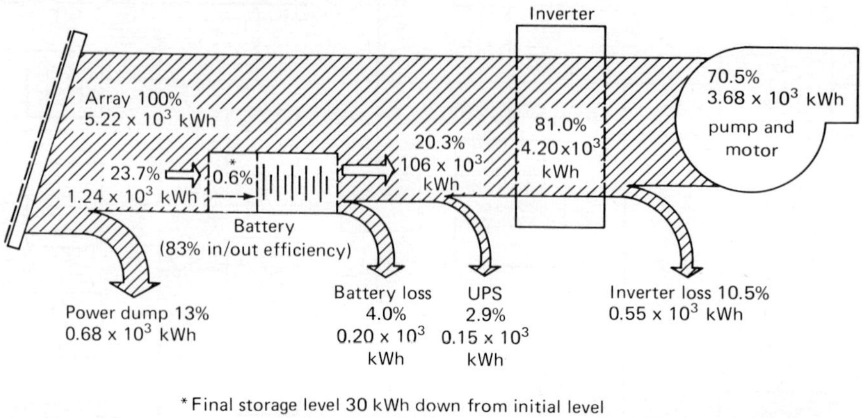

Figure A-7.4 **System energy-flow profile for the agricultural station at Mead, Nebraska, during the irrigation season of August 1 to September 10, 1978. (M.I.T. Lincoln Laboratory.)**

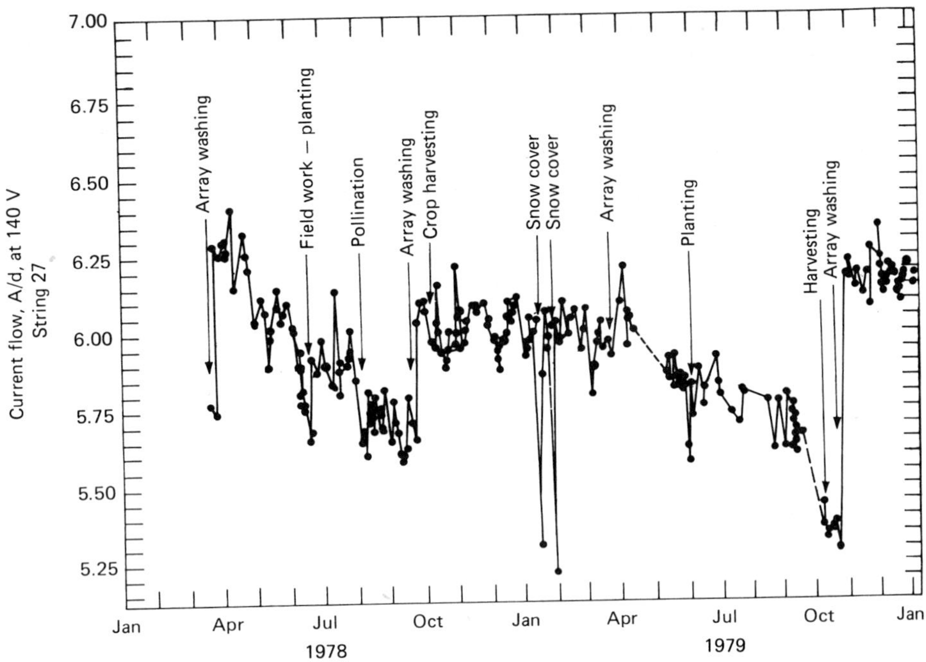

Figure A-7.5 Agricultural station system current output during 1978 and 1979. (M.I.T. Lincoln Laboratory.)

Table A-8 Remote Ranger Station System Description.

APPLICATION:	Solar-powered remote national park ranger station and residential settlement
LOCATION:	Natural Bridges National Monument, Utah
PV ARRAY OUTPUT:	100 kWp at 20°C; 220 V × 450 A (approx.)
PV ARRAY CONFIGURATION:	
ELECTRICAL:	20 branches: 5 modules in parallel by 7 in series 21 branches: 48 modules in series by 2 in parallel 14 branches: 15 modules in series by 4 in parallel 1 hybrid branch
PHYSICAL:	Rack mount: 96.7 m × 2.2 m panels in 20 rows with 4-8 panels each
PV ARRAY AREA:	Array: 1450 m^2 Land: $1^1/_3$ acre
ENERGY STORAGE:	Lead-acid battery, 750 kWh capacity 213 Vdc nominal discharge voltage
BACKUP POWER:	Diesel generator
POWER CONDITIONING:	5-kVa dc-to-ac inverter for power system control; 50 kVa inverter for site power; array-shedding battery protection circuitry
LOAD:	5 residential houses and 1 visitor center, machine shop and maintenance building, water pumping, trash compacting, external lighting
OPERATION TIME:	May 1980 to present

Figure A-8.1 **PV-powered remote ranger station and settlement (Natural Bridges National Monument). (Photo courtesy M.I.T. Lincoln Laboratory.)**

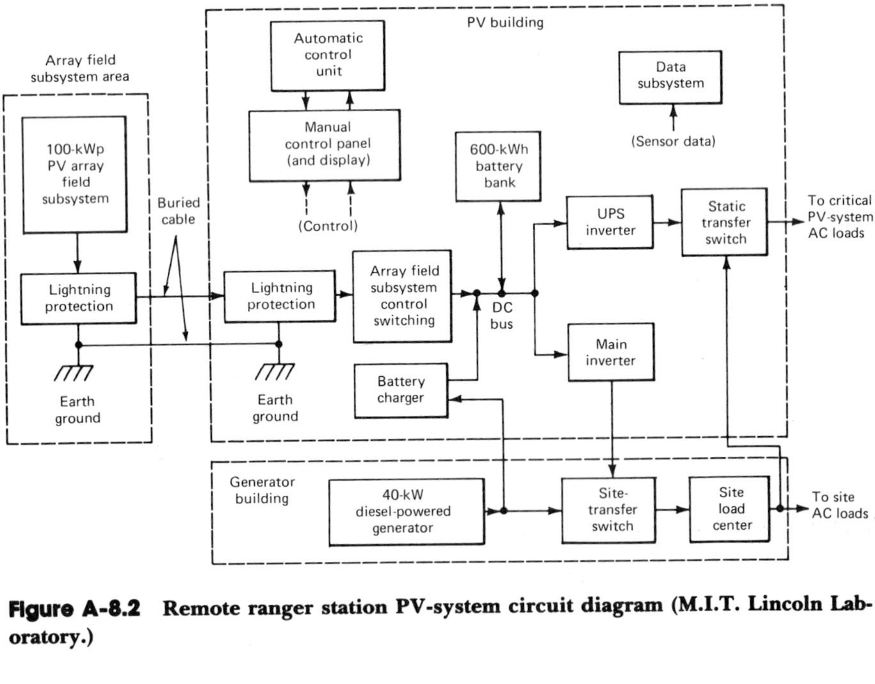

Figure A-8.2 **Remote ranger station PV-system circuit diagram (M.I.T. Lincoln Laboratory.)**

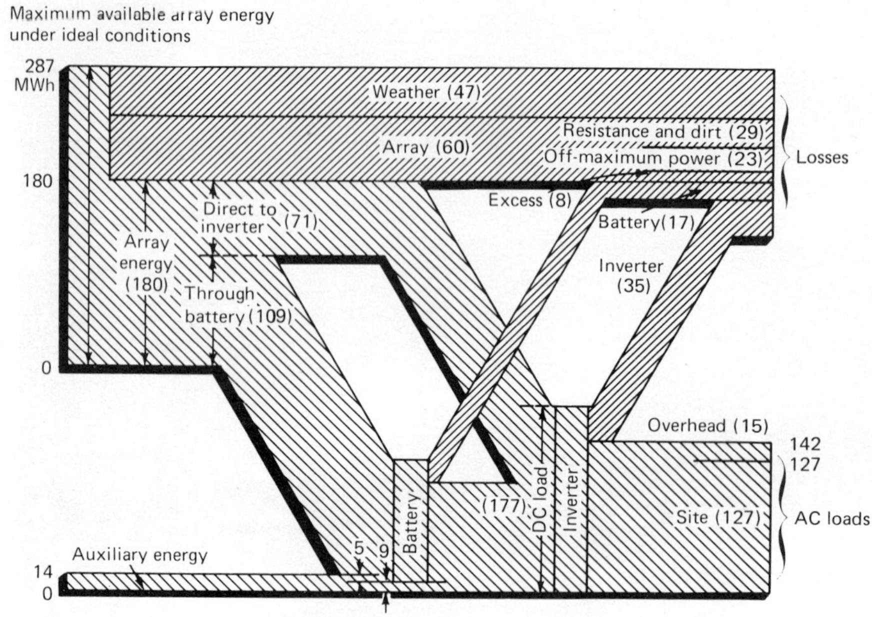

Figure A-8.3 **System energy-flow profile for a remote ranger station at Natural Bridges National Monument. All numbers are in MWh (M.I.T. Lincoln Laboratory.)**

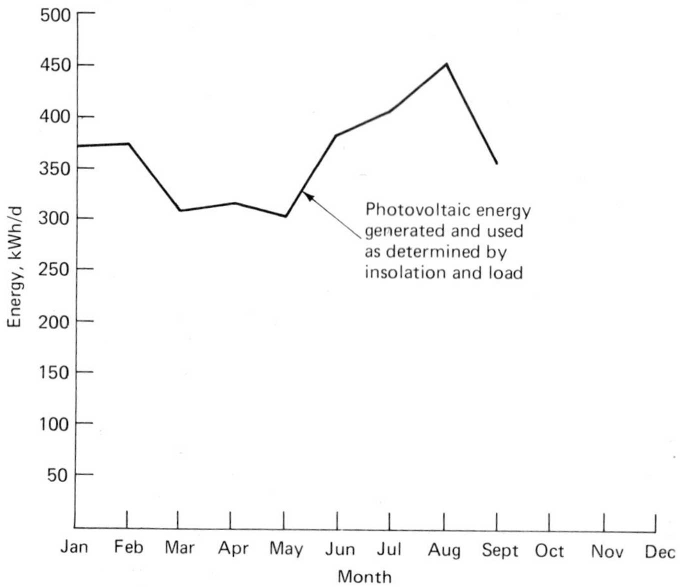

Figure A-8.4 **PV-energy output during 1981 for Natural Bridges ranger station. *Note:* The size of the battery and the load forced this system to discharge at least 20% of the available solar photovoltaic energy.**

Figure A-9 The Northeast Residential Experiment Station. (Photo courtesy M.I.T. Lincoln Laboratory.)

Table A-9.1 Northeast Residential Experiment Station Residential Photovoltaic Prototype Descriptions*

Prototype system	Module supplier	Array area, m^2	Array peak power, kWp	Mounting method	Array tilt, degrees	Inverter supplier	Inverter size, kVA
General Electric	General Electric	73.4	6.8	Direct	33.7	Abacus	6
MIT LL	Solarex	87.4	7.0	Standoff	45.0	Windworks	8
Solarex	Solarex	71.5	5.3 (est)	Standoff	40.0	Abacus	6
TriSolar Corp	Applied Solar	47.8	4.8	Integral	45.0	Windworks	8
Westing-house	ARCO Solar	74.9	5.2	Integral	45.0	Abacus	6
Carlisle House	Solarex	97.0	7.8	Standoff	45.0	Windworks	8

Source: M.I.T. Lincoln Laboratory.

*The Northeast Residential Experiment Station in Concord, Massachusetts, monitors the output performance of five residential-scale photovoltaic prototypes and the Carlisle House, and records the size and timing of electric energy used by five lived-in residences in the Greater Boston area.

TableA-9.2 Northeast Residential Experiment Station Monitored Residential Houses Descriptions.

	MH2	*MH3*	*MH4*	*MH5*	*MH6*
Occupants	5	5	4	5	4
Working adults (daytime hours)	2	2	1	2	1
Heating system	Oil	Oil and electric (2 rooms)	Oil	Oil	Oil
Domestic hot water system	Electric	Electric	Oil	Electric	Oil
Cooking	Electric	Electric	Electric	Electric	Electric
Clothes drying	Electric	Electric	Electric	Electric	Electric
Air conditioning	None	None	Window	None	Window

Source: M.I.T. Lincoln Laboratory

Table A-9.3 NE RES Month Summary of Data.

Site Location	*NERES*
Latitude:	42.46°
Longitude:	71.30°

Month: June Year: 1981

METEOROLOGICAL INFORMATION	NERES	CARLE
Average maximum ambient air temp. (°C)	26.8	N/A
Average minimum ambient air temp. (°C)	14.3	N/A
Average ambient air temp. (°C)	20.6	N/A
Average degree-days (°C days/days)	0.0	N/A
Total precipitation (cm)	7.2	—
Average wind speed (m/s)	2.5	—
Average total horizontal insolation ($kWh/M^2 \cdot day$)	5.7	—

MONITORED HOUSE (MH) INFORMATION	MH1	MH2	MH3	MH4	MH5	MH6
Average total electric energy used (kWh/day)	—	18.4	26.4	17.04	32.6	22.6
Average electric energy in use during sun hours (kWh/day)	—	11.8	20.0*	10.74	22.8	14.0

Table A-9.3 NE RES Month Summary of Data *(Continued)*.

PROTOTYPE AND ISEE PHOTVOLTAIC-ARRAY INFORMATION	MITLL	WEST	GE	TRISC	SOLEX	CARLE
Average PV DC energy (kWh/day)	31.2	20.4	28.5	20.4	—	29.2*
Average PV DC energy/rated power (kWh/day · kWp)	4.6	3.9	4.2	4.3	—	4.0*
Array peak power (kW)	6.7	4.6	6.3	4.4	—	6.2
Average array peak power (kW)	5.6	4.2	5.1	3.6	—	N/A
Average total tilt insolation: during sun hours ($kWh/m^2 \cdot day$)	4.93	4.59	5.85	4.64	—	4.47*
Average total tilt insolation: PV system on ($kWh/m^2 \cdot day$)	4.89	4.35*	N/A	4.63	—	4.45
Peak isolation (kW/m^2)	1.04	1.00	1.14	1.03	—	1.00
Average peak insolation (kW/m^2)	0.88	0.84	0.93	0.85	—	0.83
PV array efficiency (%)	7.4	6.3	N/A	9.4	—	6.7
Average maximum panel temp (°C)	57.8	51.7	63.1	47.8	—	57.2
Average minimum panel temp (°C)	12.8	10.2	11.4	11.0	—	8.9

PROTOTYPE AND ISEE POWER-CONDITIONING UNIT (PCU) INFORMATION	MITLL	WEST	GE	TRISC	SOLRX	CARLE
Average PCU AC energy output (kWh/day)	28.1	16.6	23.1*	18.4	—	26.3
Average PCU AC peak power output (kW)	5.2	3.6	4.2*	3.3	—	5.6
PCU peak power output (kWh)	6.2	3.9	5.3	4.1	—	7.0
PCU efficiency (%)	90.1	81.3	81.0*	90.1	—	90.1*
PROTOTYPE AND ISEE AC ENERGY FLOW	**MITLL**	**WEST**	**GE**	**TRISC**	**SOLRX**	**CARLE**
Average imposed load (kWh/day)	20.2	0.0	14.5	0.0	—	18.8
Average energy supplied to utility feed (kWh/day)	21.0	16.6	19.0*	18.4	—	17.6
Average energy supplied to utility feed (kWh/day)	13.4	0.0	11.0*	0.0	—	10.2
NUMBER OF DAYS OF DATA COLLECTED	25	12	16	27 0	19	

Source: M.I.T. Lincoln Laboratory.

+The load data source for the imposed load was changed from MH3 to MH4 on 6/12.

*Extimated.

Notes: MITTL had between 5 and 8 of 8 strings online. One string had current leakage problems.
WEST was down between 6/5 and 6/22 due to inverter malfunction.
GE came online 6/9.
TRISC was online full month.
SOLRX was not online.
CARLE was online full month.

Figure A-10 The Southwest Residential Experiment Station. (Photo courtesy M.I.T. Lincoln Laboratory.)

Table A-10.1 Southwest Residential Experiment Station PV Prototypes System Description.*

Prototype system	Module supplier	Array area, m^2	Array peak power, kWp	Mounting method	Array tilt, degrees	Inverter supplier	Inverter size, kVA
ARCO Solar	ARCO Solar	87.8	7.4	Direct	25	Windworks	8
ARTU	ARCO Solar	53.5	4.9	Standoff	45	Windworks	8
BDM	Motorola	54.0	4.4	Standoff	35	Abacus	6
General Electric	General Electric	74.3	6.7	Direct	26.6	Abacus	6
Solarex	Solarex	67.8	4.6	Standoff	26	Abacus	6
TEA	Motorola	49.4	4.2	Rack	26	Abacus	6
TriSolarCorp	Applied Solar	58.0	5.2	Integral	30	Windworks	8
Westinghouse	ARCO Solar	70.8	5.5	Integral	30	Abacus	6

Source: M.I.T. Lincoln Laboratory.

*The Southwest Residential Experiment Station in Las Cruces, New Mexico, monitors the output and performance of eight residential photovoltaic prototypes, and records the size and timing of electric energy used by five lived-in residences in the Las Cruces area.

Table A-10.2 Southwest Residential Experiment Station Monitored Residential Houses Descriptions.

	MH1	*MH2*	*MH3*	*MH4*	*MH5*
Occupants	5	3	4	4	2
Working adults (daytime)	1	1	2	2	Retired
Size/type of home	2060-ft^2 single-family detached	2400-ft^2 single-family detached	2650-ft^2 single-family detached	1700-ft^2 single-family detached	1600-ft^2 attached town house
Space heating system	Electric/solar	Heat pump	Electric	Natural gas	Heat pump
Cooling system	Evaporative	Heat pump	Evaporative	Evaporative	Heat pump
Domestic hot water system	Solar/electric backup	Electric	Electric	Electric	Solar/electric backup
Cooking	Electric	Electric	Electric	Electric	Electric
Clothes drying	Electric	Electric	Electric	Electric	Electric

Source: M.I.T. Lincoln Laboratory.

Table A-10.3 SW RES Monthly Summary of Data.

Site Location

Latitude: 32° 17′ N
Longitude: 106°45′ W
Elevation: 1,198 meters

Month: November Year: 1981

METEOROLOGICAL INFORMATION	SW RES
Average maximum ambient air temp. (°C)	19.66
Average minimum ambient air temp. (°C)	3.13
Average ambient air temp. (°C)	11.40
Average degree-days (°C/day)	−6.70
Total precipitation (cm)	1.47
Average wind speed (m/s)	2.06
Average total horizontal insolation (kWh/m²·day)	3.62

MONITORED HOUSE (MH) INFORMATION	MH1	MH2	MH3	MH4	MH5	
Average total electric energy used (kWh/day)	—	45.09	52.49	53.90	32.22	—
Average electric energy use: sun hours (kWh/day)	21.85	23.02	24.25	13.38	—	
NUMBER OF COMPLETE DAYS OF MH DATA*	29	26	25	29	—	

Table A-10.3 SW RES Monthly Summary of Data *(Continued).*

PROTOTYPE PHOTOVOLTAIC ARRAY INFORMATION	TSC	SX	ARTU	ARCO	TEA	BDM	GE	WST
Average PV DC energy (kWh/day)	29.31	24.14	29.49	36.66	—	—	33.76	30.28
Average PV DC energy/rated power (kWh/day·kWp)	5.64	5.25	6.02	4.95	—	—	5.04	5.50
Maximum array peak power (kW)	5.12	4.50	4.89	6.78	3.84	4.04	5.69	4.95
Average array peak power (kW)	4.62	3.95	4.34	5.97	—	—	5.30	4.57
Average total tilt insolation: sun hours (kWh/m^2·day)	5.78	5.40	6.12	5.27	5.42	5.74	5.42	5.81
Insolation during system on hours/total insolation (%)	99.87	96.31	99.70	98.65	—	—	98.85	98.80
Maximum peak insolation (kW/m^2)	1.04	1.01	1.09	1.00	1.00	1.06	0.99	1.04
Average peak insolation (kW/m^2)	0.91	0.85	0.95	0.85	0.86	0.90	0.87	0.91
PV array efficiency (%)	8.26	6.15	8.26	7.31	—	—	7.66	6.79
Average maximum panel temp (°C)	44.23	55.47	50.80	53.68	48.45	53.89	58.18	50.09
Average minimum panel temp (°C)	0.19	−0.45	1.36	−1.49	−0.44	1.00	−0.94	1.64

PROTOTYPE POWER CONDITIONING UNIT (PCU) INFORMATION	TSC	SX	ARTU	ARCO	TEA	BDM	GE	WST
Average PCU AC energy output (kWh/day)	27.12	20.87	26.01	32.66	—	—	28.67	25.69
Average PCU AC peak power output (kW)	4.29	3.45	3.86	5.36	—	—	4.55	3.92

Maximum PCU AC peak power output (kW)	4.77	3.88	4.43	6.06	3.25	3.60	4.86	4.22
PCU efficiency (%)	92.91	87.20	88.73	89.70	—	—	85.42	85.41
PROTOTYPE AC ENERGY FLOW	TSC	SX	ARTU	ARCO	TEA	BDM	GE	WST
Average imposed load 5 AM to 8 AM (kWh/day)**	20.58	18.46	20.04	24.84	—	—	19.51	20.08
Average energy supplied to utility feed (kWh/day)	18.52	13.95	17.58	23.00	—	—	20.68	17.51
Average energy supplied to utility feed (kWh/day)	12.77	11.48	12.15	15.92	—	—	12.03	12.57
NUMBER OF COMPLETE DAYS OF PROTOTYPE DC AND AC DATA+	19	23	19	25	5	5	16	26
NUMBER OF COMPLETE DAYS OF PROTOTYPE LOAD, UTILITY DATA†	24	23	26	25	8	0	16	26

Source: M.I.T. Lincoln Laboratory.

+*A complete day is defined as more than 23 hours of data collected.

**Load imposed by MH4.

+A complete day is defined as a full day of DC and AC data available for comparison.

†A complete day is defined as a full day of load, energy to utility, and energy from utility data available for comparison.

Notes: No data are included from 24 November due to a systemwide test.
TSC was online all month. It had some nighttime load problems and invalid DC data.
SX was online all month. It had some nighttime load and instrumentation problems.
ARTU was online all month. It had some nighttime load problems and invalid DC data.
ARCO was online until Gemini inverter failure 30 November. It had some nighttime load problems.
TEA was offline 3–20 November. Database was not sufficient for meaningful comparison; data were deleted.
BDM was offline until 22 November. Database was not sufficient for meaningful comparison; data were deleted.
GE was offline 3–9 and 12–13 November due to Abacus inverter failure. There were also some data transmission problems.
WST was online all month.

APPENDIX B

HANDBOOK FOR VARIOUS BLOCK III AND BLOCK IV SILICON SOLAR-CELL MODULES*

The block categories were set up by the U.S. Department of Energy to establish minimum criteria for each major government purchase of photovoltaic modules. Between the years 1970 and 1982, the government developed five progressively more stringent block standards: Blocks I to V. The major Block III performance criteria are:

1. The configuration must permit assembly of modules into a 4-ft-square subarray that produces not less than 60 W at rated output voltage when operated at a cell temperature of 60°C and exposed to insolation through an optical air mass of AM1.5 at a module irradiance of 100 mW/cm^2.
2. Breakdown voltage from terminals to ground must exceed 1500 Vdc.
3. Insulation resistance from terminals to ground, when measured at 1000 Vdc, must not be less than 100 MΩ.
4. The modules must withstand the following test environment:
   a. 50 thermal cycles between −40°C and +90°C.
   b. 5 cycles of 95% relative humidity between 23°C and 41°C.
   c. 100 cycles of simulated wind loading between +50 lb/ft^2 and −50 lb/ft^2.

Source: M. I. Smokler, *User Handbook for Block III Silicon Solar Cell Modules,* Low Cost Solar Array Project, JPL Document Jet Propulsion Laboratory, 1979; *User Handbook for Block IV Silicon Solar Cell Modules,* Low Cost Solar Array Project, JPL Document, Jet Propulsion Laboratory, 1982.

Table B-1 Characteristics for Five Block III PV Modules

IDENTIFYING Manufacturer	ARCO SOLAR	MOTOROLA	SENSOR TECHNOLOGY	SOLAR POWER	SOLAREX
Mfr's model no.	10699-C	P-0170-770-J	20-10-1646	E10008-F	A·0221-G
PHYSICAL					
Photographic views	Figure B-1	Figure B-1	Figure B-1	Figure B-1	Figure B-1
Overall dimensions (inches)					
Length	46.0	23.0	22.9	46.0	22.9
Width	9.1 23.0	11.3	15.3	22.9	
Height	1.4	2.0	1.8	1.9	1.8
Weight (lbs)	8.1	14.5	8.2	16.3	9.8
No. of modules per 4-foot-square subarray	5	4	8	3	4
Subarray planarity Romt (in/ft)	0.25	0.25	0.25	0.25	0.25
Module drawing and materials	Figure B-5	Figure B-3	Figure B-7	Figure B-9	Figure B-11
Electrical connections	Figure B-5	Figure B-3	Figure B-7	Figure B-9	Figure B-11
No. of cells	41	48	44	40	42
Cell diameter (inches)	3	3	2.2	4	3
Cell configuration	N·P	N·P	N·P	P·N	N·P

ELECTRICAL					
Qualification performance[a]					
Power rated (watts)	18.5	23.2	9.4	29.5	18.5
Voltage rated (volts)	15.8	5.0	16.5	15.8	15.8
Current (amps)	1.17	4.64	0.57	1.87	1.17
SOC performance (sample)[b]					
Power maximum (watts)	19.3	23.1	10.1	31.4	19.1
Voltage at max. power (volts)	15.6	5.1	18.3	16.7	15.7
Current at max. power (amps)	1.24	4.52	0.55	1.88	1.21
Voltage open circuit (volts)	20.9	6.4	22.5	21.6	21.3
Current short circuit (amps)	1.40	4.91	0.62	1.98	1.42
Fill factor	0.66	0.74	0.72	0.73	0.63
Efficiency module (%)	7.2	6.8	6.1	6.9	5.6
Efficiency encapsulated cell (%)	10.4	10.6	9.8	9.7	9.8
Temperature characteristics (I/V curves)	Figure B-6	Figure B-4	Figure B-8	Figure B-10	Figure B-12
Circuit diagram	Figure B-5	Figure B-3	Figure B-7	Figure B-9	Figure B-11
Insulation resistance, min (Ω)	100	100	100	Not applicable	100
Breakdown voltage, min. (Vdc)[c]	1500	1500	1500	Not applicable	1500

Table B-1 Characteristics for Five Block III PV Modules *(Continued)*

IDENTIFYING Manufacturer	ARCO SOLAR	MOTOROLA	SENSOR TECHNOLOGY	SOLAR POWER	SOLAREX
Mfr's model no.	10699-C	P-0170-770-J	20-10-1646	E10008-F	A·0221-G
THERMAL					
Nominal operating cell temperature (°C)[d]	57.2	61.0	49.0	52.5	63.1
ENVIRONMENTAL					
Temperature range (°C)[e]	−40 to +90	−40 to +90	−40 to +90	−40 to +90	−40 to +90
Humidity maximum relative (%)[e]	95	95	95	95	95
Wind load maximum (lb/ft^2)[e]	±50	±50	±50	±50	±50
LIFE					
Service life design goal (yrs)	10	10	10	10	10

Source:

[a]Each module was required to produce not less than 86% of rated power when loaded to provide rated voltage under the following conditions:

a. Module irradiated with 100 mW/cm^2 insolation at air mass 1.5 (AM1.5).

b. Cell temperature equal to 60°C.

[b]Standard operating conditions (SOC) are:

a. Module irradiated with 100 mW/cm insolation at air mass 1.5 (AM1.5).

b. Cell temperature equal to NOCT (see note *d*).

The data presented here for each module design show the characteristics obtained by measurement and extrapolation of the performance of one sample module of that design.

[c]Modules should not be series-connected to produce system operating voltage in excess of 250 Vdc at standard operating conditions (SOC).

[d]Nominal operating cell temperature (NOCT) is the cell temperature with the module in the standard thermal environment (STE), defined as follows:

Insolation	= 100 mW/cm^2
Air temperature	= 200°C
Average wind velocity	= 1 m/s
Electrical load	= open circuit
Mounting	= normal to solar noon on structure typical of application

[e]The data given are the test limits, not measured limits.

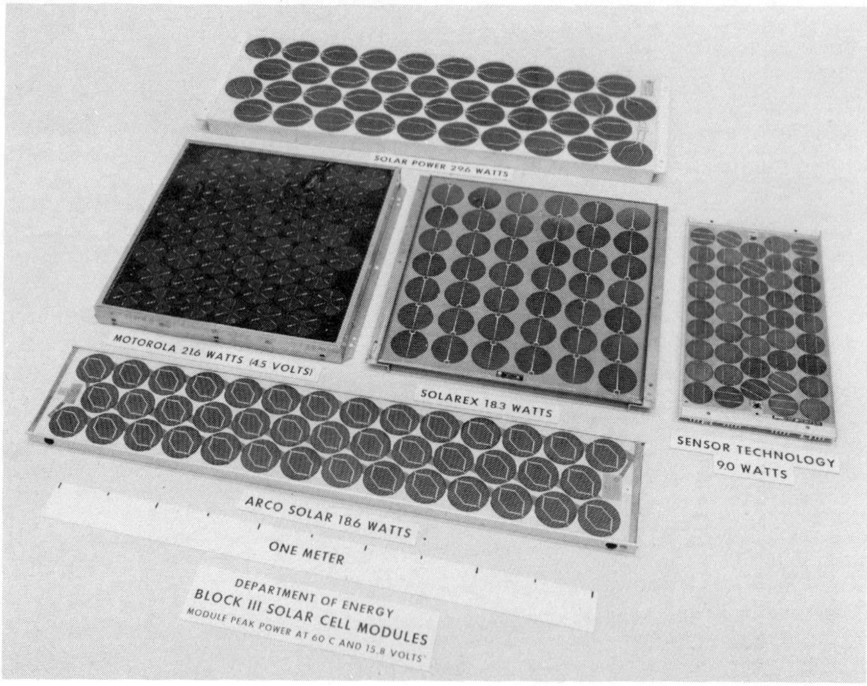

Figure B-1 Four Block III PV modules. (Photo courtesy Jet Propulsion Laboratory.)

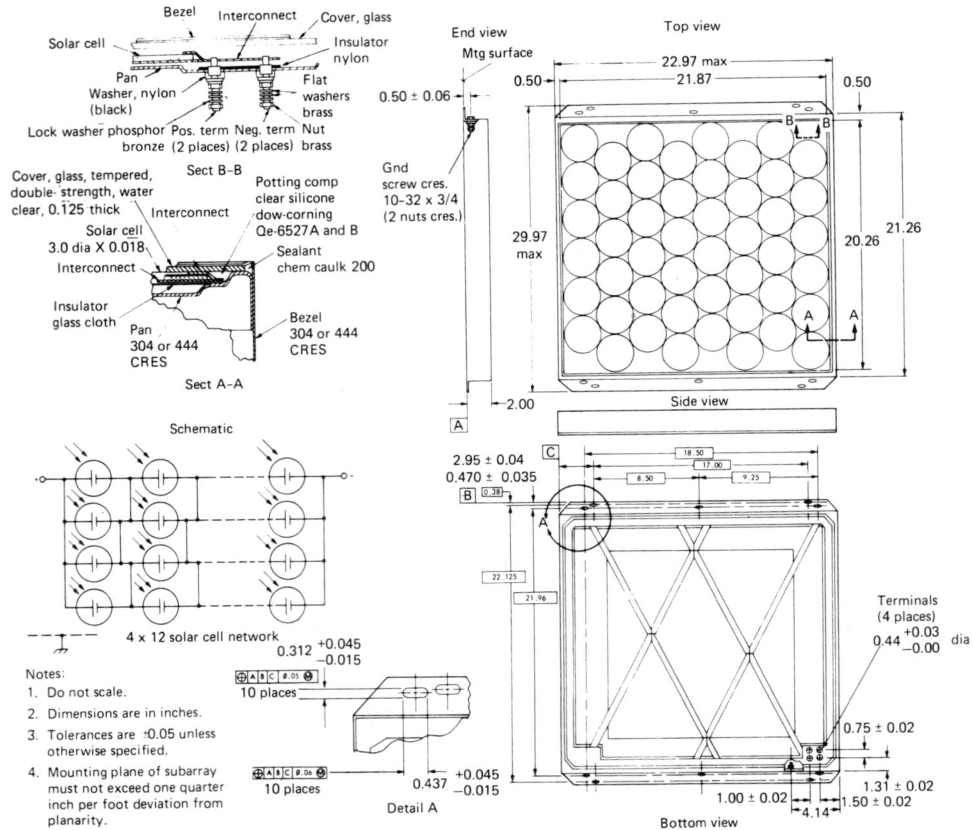

Figure B-2 Motorola module: drawing. (Jet Propulsion Laboratory.)

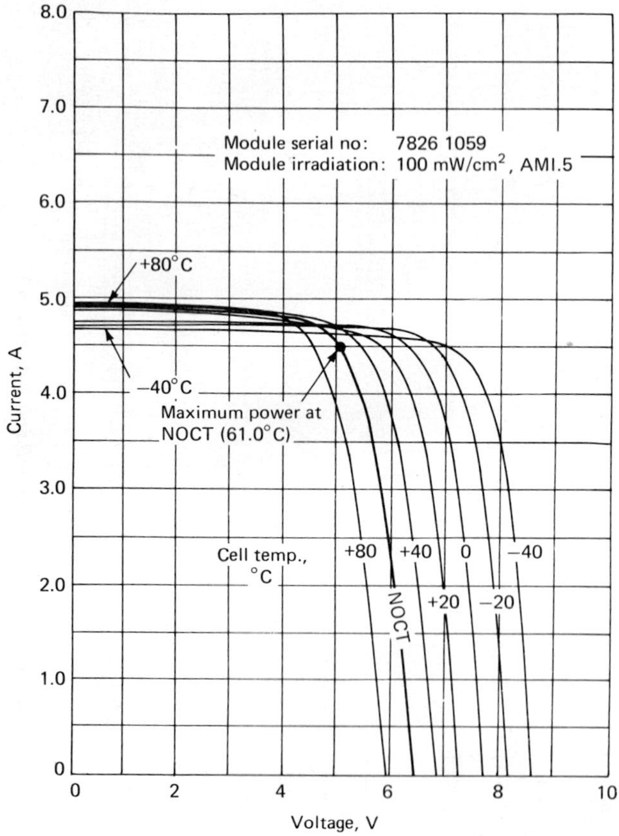

Figure B-3 Motorola module: *I-V* curves. (Jet Propulsion Laboratory.)

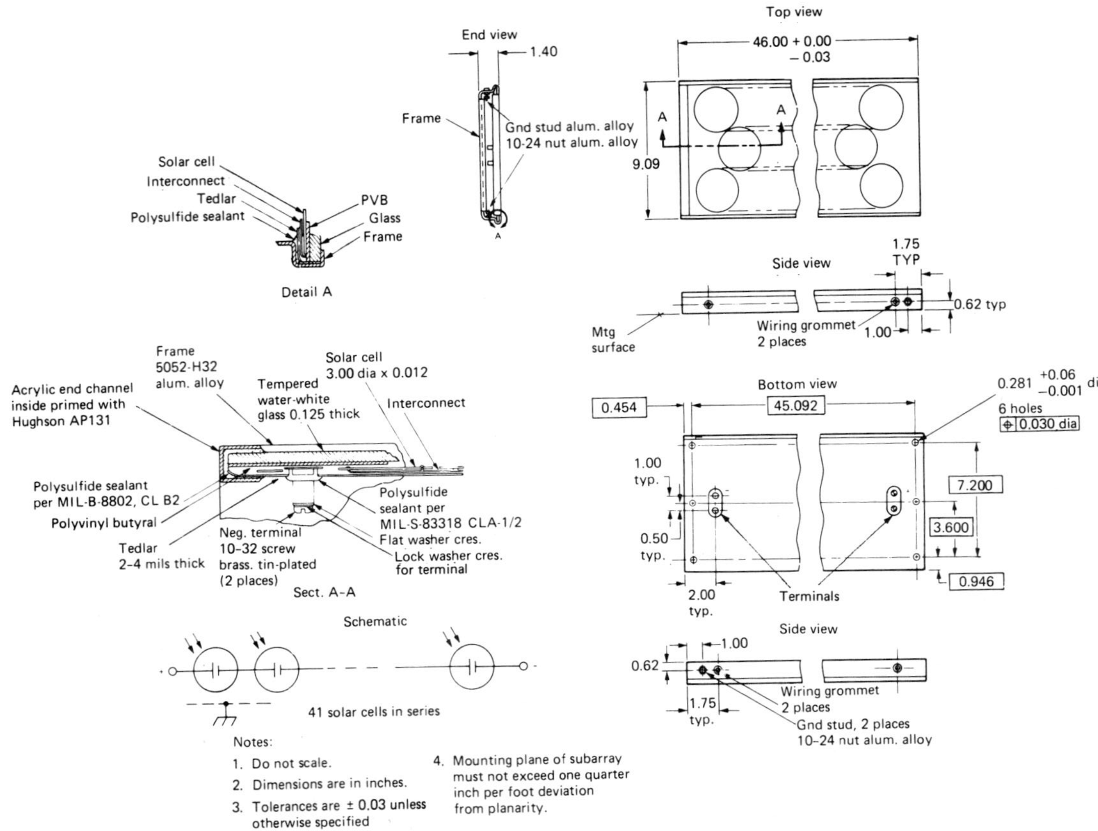

Figure B-4 ARCO Solar module: drawing. (Jet Propulsion Laboratory.)

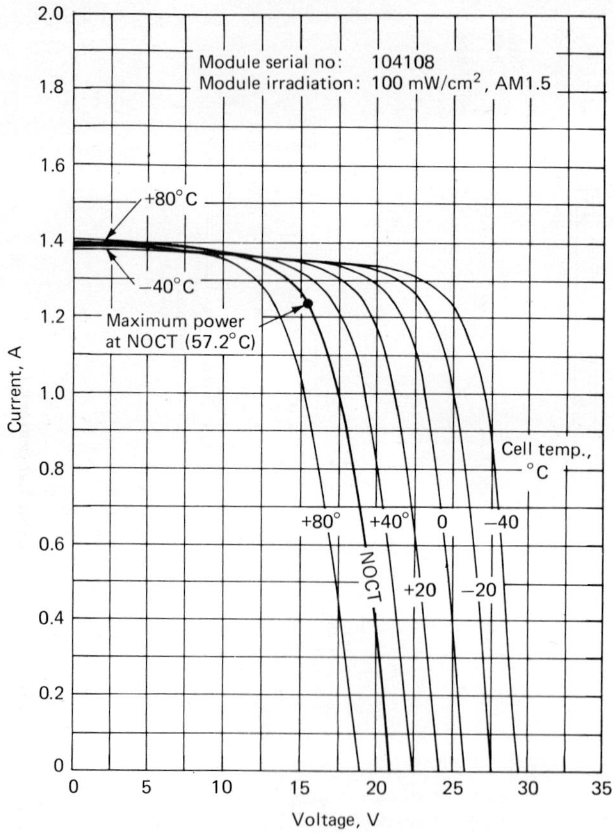

Figure B-5 ARCO Solar module: *I-V* curves. (Jet Propulsion Laboratory.)

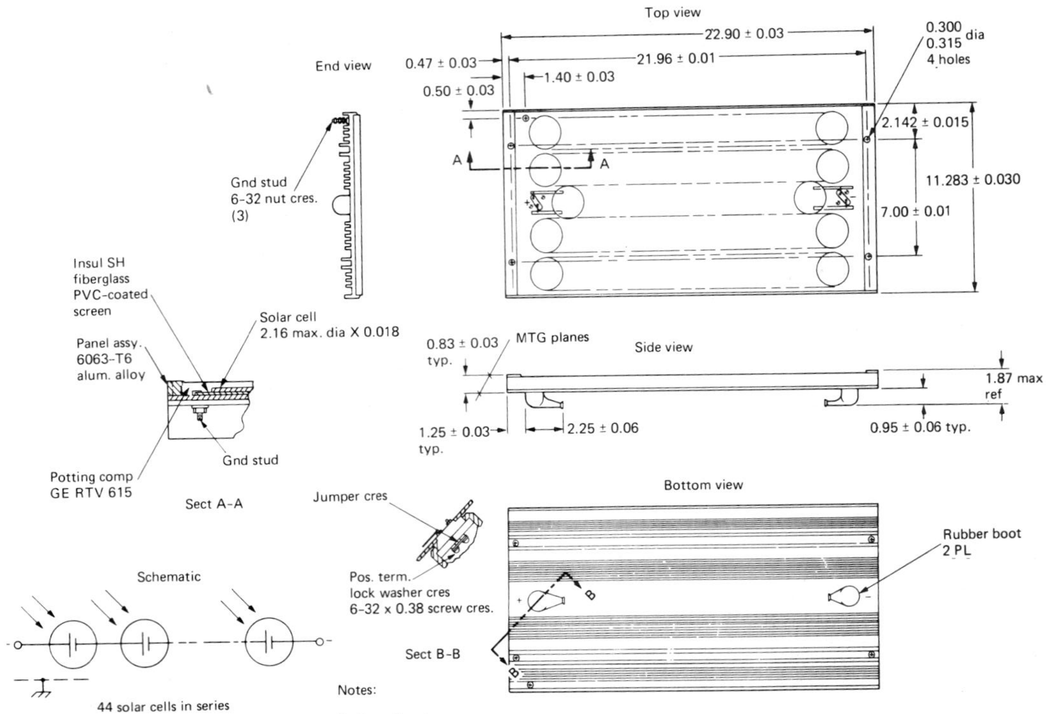

Notes:

1. Do not scale.
2. All dimensions are in inches.
3. Mounting plane of subarray must not exceed one quarter inch per foot deviation from planarity.

Figure B-6 **Sensor Technology module: drawing. (Jet Propulsion Laboratory.)**

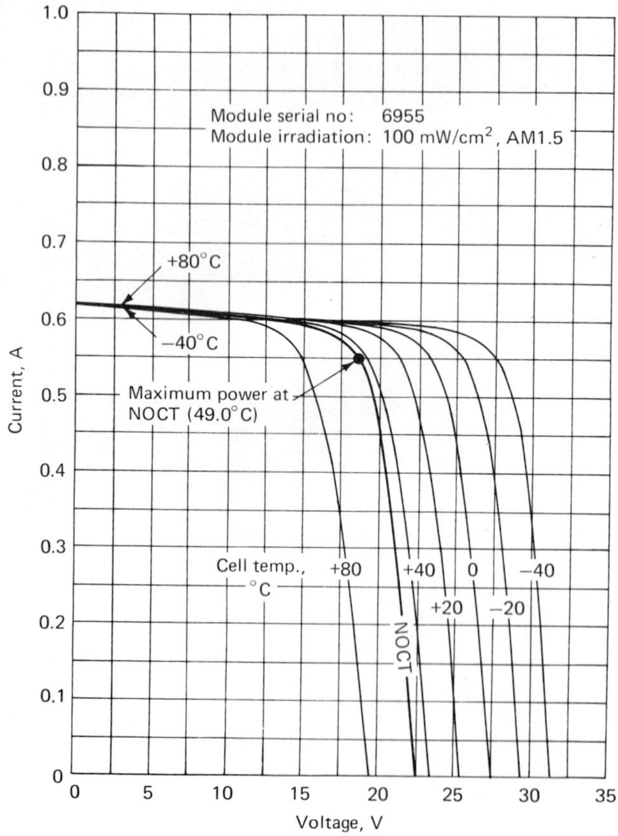

Figure B-7 **Sensor Technology module: *I-V* curves. (Jet Propulsion Laboratory.)**

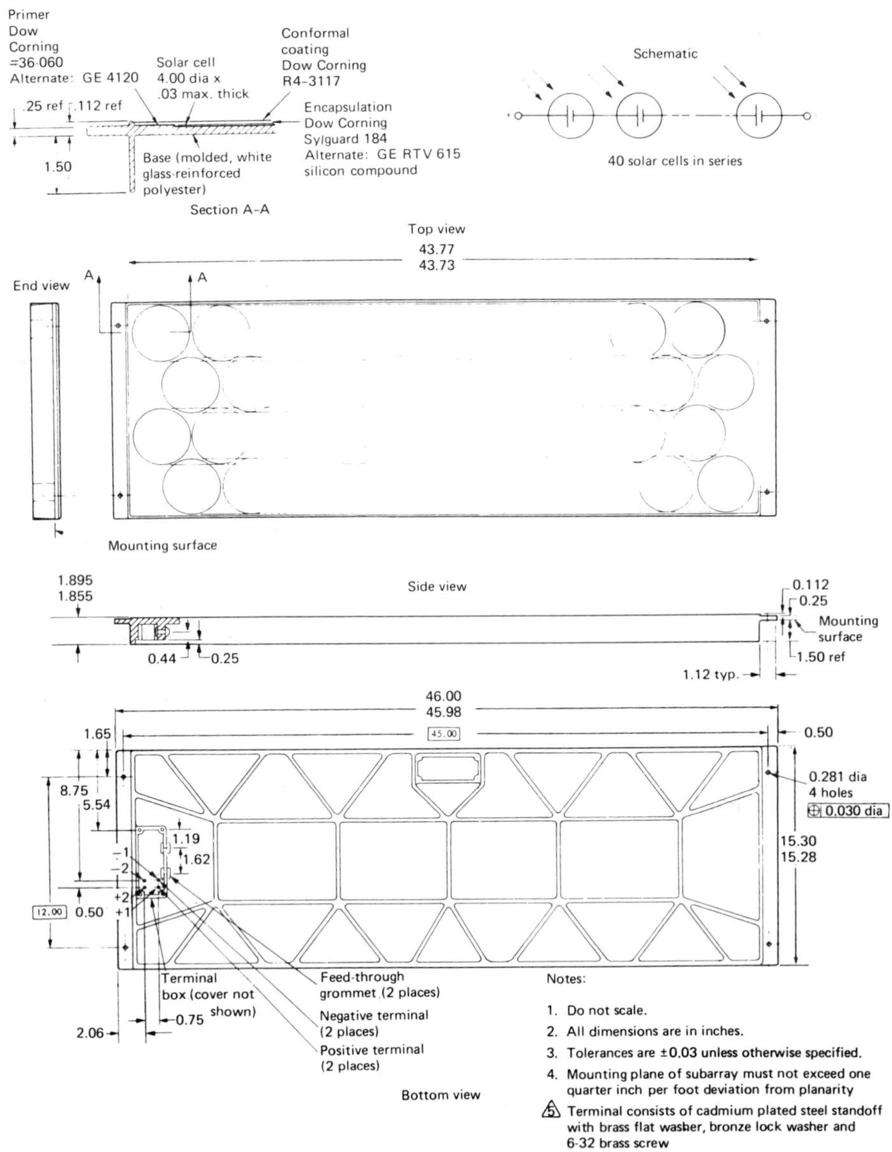

Figure B-8 **Solar Power module: drawing. (Jet Propulsion Laboratory.)**

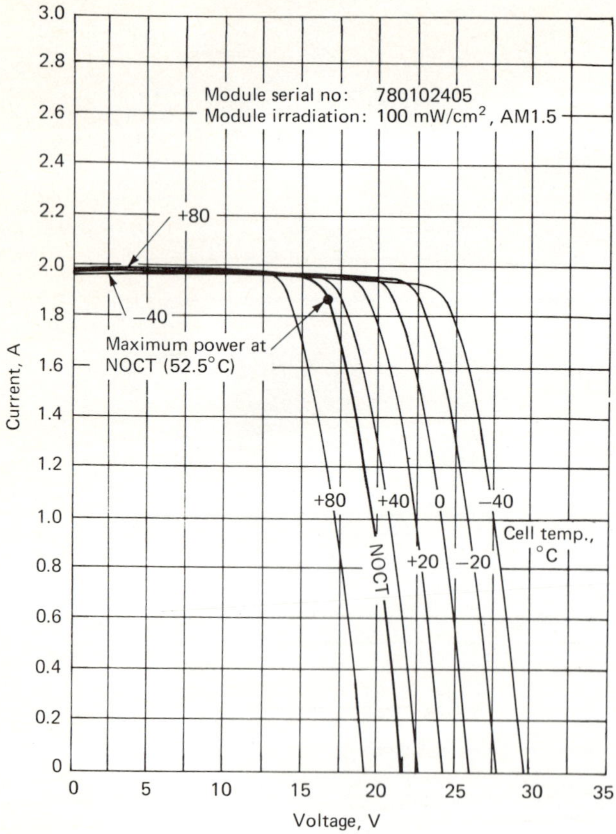

Figure B-9 Solar Power module: *I-V* curves. (Jet Propulsion Laboratory.)

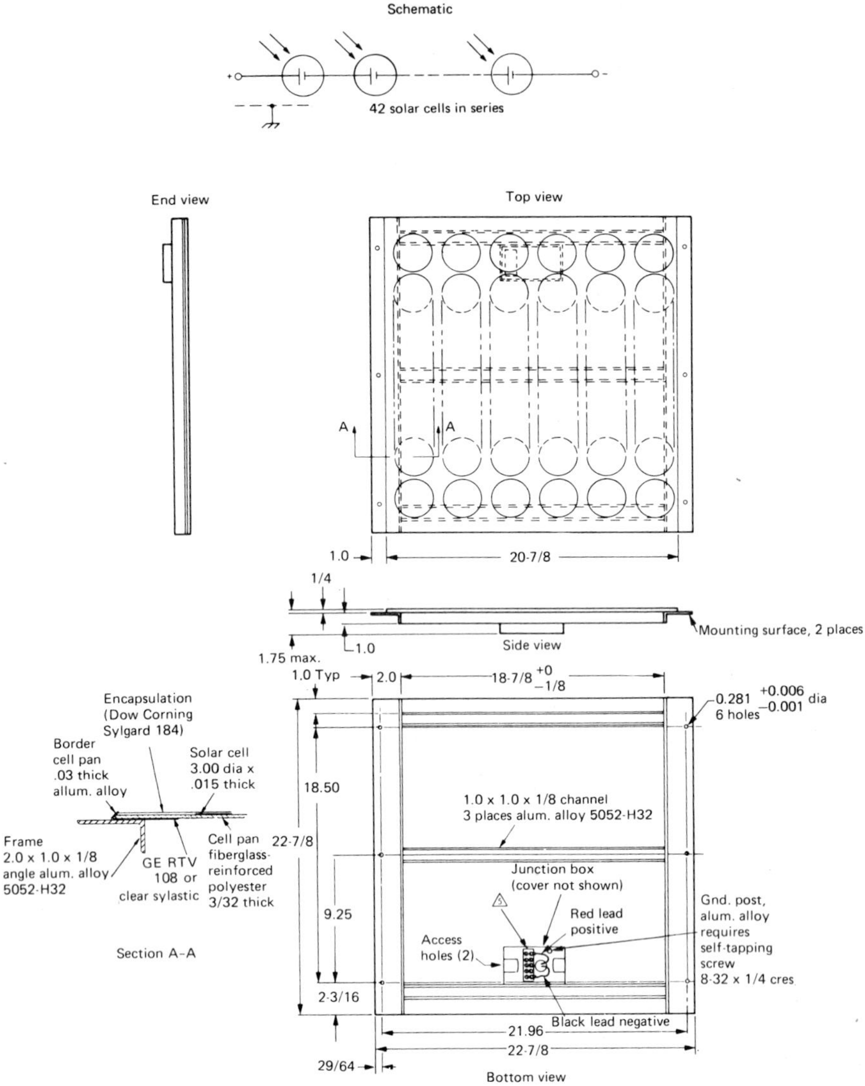

Notes:

1. Do not scale.
2. All dimensions are in inches.
3. Unless otherwise specified tolerances are: fractions: ±1/32, decimals: ±0.020
4. Mounting plane of subarray must not exceed one quarter inch per foot deviation from planarity.
5. 5 terminal, closed back, terminal block-670, screw

Figure B-10 **Solarex module: drawing. (Jet Propulsion Laboratory.)**

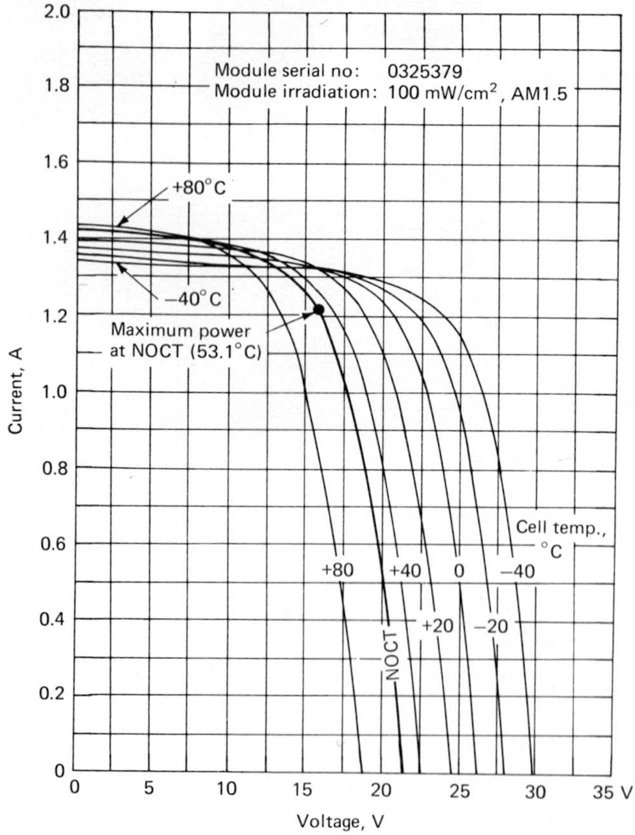

Figure B-11 Solarex module: *I-V* curves. (Jet Propulsion Laboratory.)

Table B-2 Block IV Module Characteristics

Manufacturer	ARCO Solar	ASEC	GE	Motorola	Photowatt	Solarex	Solarex	Spire
IDENT								
MFR's Part No.	012110-E	60-3062-F	47J254977G1-C	MSP43D40-G	ML-1961-D	580-BT-L-C	580-BT-R-C	058-0007-A
Module Type[a]	INT	INT	RES	INT	INT	INT	RES	INT
PHYSICAL								
Overall dimensions, mm (in)								
Length	1219 (48.0)	1198 (47.2)	818 (32.2)	1198 (47.2)	1199 (47.2)	1200 (47.3)	1193 (47.0)	1200 (47.2)
Width	305 (12.0)	696 (27.4)	669 (26.4)	356 (14.0)	444 (17.5)	635 (25.0)	628 (24.7)	417 (16.4)
Height	54 (2.1)	39 (1.5)	7.6 (0.3)	38 (1.5)	51 (2.0)	51 (2.0)	16 (0.63)	62 (2.4)
Weight, kg (lbs)	5.2 (11.4)	13.5 (29.7)	4.0 (8.8)	5.8 (12.8)	7.4 (16.3)	13.9 (30.6)	11.2 (24.6)	7.8 (17.2)
Support structure planarity rqmt,[b] mm/m (in/ft)	20 (0.25)	20 (0.25)	20 (0.25)	20 (0.25)	20 (0.25)	20 (0.25)	20 (0.25)	20 (0.25)
Cells								
Quantity	35	136	19	33	72	72	72	108
Size, mm (in)	Dia: 102.9 (4.05)	Dia: 76.2 (3.0)	Dia: 100 (3.94)	100.3 × 100.3 (3.95 × 3.95)	Dia: 76.2 (3.0)	95 × 95 (3.74 × 3.74)	95 × 95 (3.74 × 3.74)	63.5 × 63.5 (2.5 × 2.5)
Packing factor	0.76	0.74	0.76	0.76	0.62	0.85	0.87	0.84
Base material	Cz	Cz	Cz	Cz	Cz	SEMI-XTL	SEMI-XTL	Cz
Junction	N/P	N/P	N/P	N/P,P$^+$	N/P,P$^+$	N/P,P$^+$	N/P,P$^+$	N/P,P$^+$
Front metallization	Printed Ag	Ti-Pd-Ag	Printed Ag	Pd-Ni-solder	Pd-Ni-solder	Ti-Pd-Ag	Ti-Pd-Ag	Ti-Pd-Ag
Back metallization	Printed Al, Ag	Ti-Pd-Ag	Printed Al, Ag	Pd-Ni-solder	Ni-solder	Al-SnAg	Al-SnAg	Ti-Pd-Ag

Table B-2 Block IV Module Characteristics *(Continued)*

Manufacturer	ARCO Solar	ASEC	GE	Motorola	Photowatt	Solarex	Solarex	Spire
ELECTRICAL								
Nominal Performance[c]								
Power, rated, W	32.0	71.0	14.4	32.9	33.0	53.0	50.0	49.1
Voltage, rated, V	15.0	14.0	6.6	15.0	5.0	13.5	4.2	14.0
Current, A	2.13	5.07	2.18	2.19	6.6	3.9	11.9	3.5
SOC Performance[c,d,e]								
Power maximum, W	31.5	74.7	14.2	33.1	33.4	54.8	50.9	48.2
Voltage at max power, V	14.6	14.4	6.6	14.5	4.85	13.6	4.31	13.5
Current at max power, A	2.16	5.18	2.15	2.28	6.88	4.02	11.8	3.56
Voltage, open circuit, V	19.2	18.3	9.3	17.8	6.24	17.5	5.74	18.0
Current, short circuit, A	2.43	5.46	2.53	2.53	7.63	4.61	13.7	3.68
Fill factor	0.70	0.75	0.60	0.74	0.70	0.70	0.65	0.72

Efficiency, module, %	8.5	9.0	7.3	7.8	6.3	7.2	6.8	9.6
Eff., encapsulated cell, %	11.2	12.2	9.6	10.3	10.2	8.5	7.8	11.4
28°C Performance[e]								
Power maximum, W	35.7	84.6	18.8	37.3	38.6	62.6	61.2	57.0
Voltage at max power, V	16.6	16.5	8.5	16.2	5.68	16.1	5.27	16.2
Current at max power, A	2.15	5.11	2.21	2.30	6.79	3.90	11.6	3.52
Voltage, open circuit, V	21.0	20.2	11.0	19.5	6.98	19.6	6.70	20.3
Current, short circuit, A	2.42	5.40	2.53	2.50	7.58	4.50	13.2	3.64
Fill factor	0.70	0.78	0.68	0.76	0.73	0.71	0.69	0.77
Efficiency, module, %	9.6	10.1	9.6	8.8	7.2	8.2	8.2	11.4
Eff., encapsulated cell %	12.6	13.6	12.6	11.6	11.6	9.6	9.4	13.6
Breakdown voltage, min.,[b,f] Vdc	2000	2000	note *g*	2000	2000	2000	note *g*	2000
THERMAL								
NOCT,[d] °C								
At 100 mW/cm² insolation	52	54	68	56	54	56	65	56
At 80 mW/cm² insolation	46	47	58	49	47	49	56	49
Temperature coefficient[h]								
$\Delta V/\Delta T$ (V/°C)	−0.086	−0.083	−0.049	−0.078	−0.030	−0.086	−0.029	−0.097

Table B-2 Block IV Module Characteristics *(Continued)*

Manufacturer	*ARCO Solar*	*ASEC*	*GE*	*Motorola*	*Photowatt*	*Solarex*	*Solarex*	*Spire*
ENVIRONMENT								
Temperature range,[b] (°C)	−40 to +90	−40 to +90	−40 to +90	−40 to +90	−40 to +90	−40 to +90	−40 to +90	−40 to +90
Humidity, maximum relative,[b] %	90	90	90	90	90	90	90	90
Wind load, maximum,[b] kPa (lb/ft^2)	±2.4 (±50)	±2.4 (±50)	±1.7 (35)	±2.4 (±50)	±2.4 (±50)	±2.4 (±50)	±1.7 (±35)	±2.4 (±50)
Hail impact, maximum hailstone,[b] mm (in)	20.0 (0.75)	20.0 (0.75)	20.0 (0.75)	20.0 (0.75)	20.0 (0.75)	20.0 (0.75)	20.0 (0.75)	20.0 (0.75)

Source: M. I. Smokler, *User Handbook for Block IV Silicon Solar Cell Modules*, JPL Document No. 5101-214, Jet Propulsion Laboratory, Pasadena, Calif., 1982.

[a]INT = Module intended for use in intermediate-load centers, defined here as installations providing 20 to 500 kW.
RES = Module intended for use on a single-family residence in installations providing 2 to 10kW.

[b]The data given are tested limits, not module limits.

[c]Each module is expected to produce not less than 90% of rated power when loaded to provide rated voltage under Standard Operating Conditions (SOC), i.e.:

1. Module irradiated with 100 mW/cm^2 insolation at air mass 1.5 (AM1.5) spectrum.
2. Cell temperature equal to NOCT per Block IV Specifications; see note *d*).

[d]The Block IV Specifications define NOCT (nominal operating cell temperature) as the cell temperature with the module in the standard thermal environment defined as follows:

Insolation	= 100 mW/cm^2
Air temperature	= 20°C
Average wind velocity	= 1 m/s
Electrical load	= open circuit
Mounting	= normal to solar noon on structure typical of application

Practice at the time of publication is to measure NOCT with the module in the nominal thermal environment, which is the same as the standard thermal environment except that the insolation level is 80 mW/cm^2.

[e]The data presented here for each module design were obtained by measurement and extrapolation of the performance of one sample module of that design. The radiation source was a large-area pulsed solar simulator calibrated by use of a calibrated reference cell of the same spectral response as the module to irradiate the module with the equivalent of 100 mW/cm^2 at AM1.5. Module temperature was approximately 20°C. Extrapolation was performed by computer, based upon a set of measured temperature coefficients (voltage, current, and series resistance) for each module design.

[f]Modules should not be series-connected to obtain system voltages under worst-case conditions (100 mW/cm^2 insolation, 0°C cell temperature, open circuit) exceeding:

500 V for intermediate load modules
250 V for residential modules

[g]Not applicable. This module has no exposed conductive surface and is intended for installation in a nonconductive assembly.

[h]These coefficients are for use in the neighborhood of the maximum power points on the module *I-V* curves. They are useful for determining power output at a selected voltage and temperature when the available *I-V* curve was made at a different temperature.

Figure B-12 (*a*) **Four Block IV PV modules** ***(left to right)*****: ARCO Solar (32 Wp), ASEC (71 Wp), Motorola (33 Wp), and Photo watt (26 Wp).** (*b*) **Three Block IV PV modules** ***(left to right)*****: Solarex (56 Wp), Solar Power (56 Wp), and SPIRE (50 Wp). (Photos courtesy Jet Propulsion Laboratory.)**

APPENDIX C
CLIMATIC DATA FOR 20 SITES WITHIN THE CONTINENTAL UNITED STATES

Clearness Index, Dry-Bulb Daytime and Daily Average Temperature, Wind Speed, and Insolation on a Horizontal and Tilted Surface*

United States Sites

Astoria	OR	Fargo	ND
Boise	ID	Grand Junction	CO
Boston	MA	Kansas City	MO
Caribou	ME	Kingsville	TX
Casper	WY	Norfolk	VA
Charleston	SC	Oakland	CA
Chattanooga	TN	Orlando	FL
Cleveland	OH	Philadelphia	PA
El Paso	TX	Phoenix	AZ
El Toro	CA	Rochester	NY

Source: Y. P. Gupta and S. K. Young, *Design Handbook for Photovoltaic Power Systems,* Sandia Laboratory, Albuquerque, N. Mex., 1981, Volume II.

NOAA Technical Memorandum ERL ARL-77, *Solar Radiation and Climatic Data for Quasi-Homogeneous Regions of the United States,* 1979.

For additional climatic data, consult the above source or the *Insolation Data Manual,* Solar Energy Research Institute, Golden, Colo., 1980 (SERI/SP-755-789).

*Horizontal to tilted insolation calculations performed using Collares-Pereira and Rabel adaptions of Lui and Jordan's equations as written up in *Solar Energy,* **23,** 1979.

Table C-1 Climatic Data for Astoria OR Latitude 46.2°; Longitude 123.8°

Climate statistics

	Month												
	J	F	M	A	M	J	J	A	S	O	N	D	Year
Cleanliness index *k* (daily average) (dimensionless)	0.316	0.369	0.397	0.435	0.470	0.446	0.495	0.486	0.484	0.416	0.342	0.305	0.414
Dry bulb													
(daytime average) °C	6.3	7.7	8.0	9.8	12.4	14.7	16.6	16.9	16.3	13.1	9.6	7.2	11.6
(daily average) °C	5.4	6.5	6.8	8.6	11.3	13.6	15.4	15.7	14.7	11.6	8.3	6.3	10.3
Wind speed (daily average) m/s	4.2	3.9	4.0	3.8	3.8	3.7	3.8	3.4	3.3	3.2	3.7	4.0	3.7

Insolation on collector, kWh/m²·d

Collector type	Collector tilt	Month												
		J	F	M	A	M	J	J	A	S	O	N	D	Year
Flat plate—south facing	Horizontal	0.99	1.72	2.73	3.95	5.07	5.13	5.51	4.73	3.73	2.25	1.22	0.82	3.15
	30 degrees	1.64	2.40	3.23	4.19	4.99	4.88	5.33	4.87	4.29	2.98	1.92	1.47	3.52
	35 degrees	1.72	2.47	3.26	4.15	4.88	4.75	5.20	4.81	4.30	3.05	2.00	1.55	3.52
	40 degrees	1.79	2.53	3.28	4.10	4.76	4.60	5.05	4.72	4.30	3.10	2.07	1.62	3.50
	45 degrees	1.85	2.58	3.27	4.03	4.61	4.43	4.87	4.61	4.27	3.14	2.12	1.68	3.46
	50 degrees	1.89	2.60	3.25	3.93	4.44	4.24	4.68	4.48	4.21	3.15	2.17	1.73	3.40
	55 degrees	1.93	2.62	3.21	3.82	4.25	4.03	4.46	4.33	4.14	3.15	2.19	1.76	3.33
	60 degrees	1.95	2.61	3.15	3.69	4.04	3.81	4.23	4.15	4.04	3.13	2.21	1.79	3.24
	Vertical	1.82	2.28	2.47	2.57	2.47	2.19	2.48	2.74	3.03	2.64	2.02	1.71	2.37

Table C-2 Climatic Data for Boise ID Latitude 43.6°; Longitude 116.2°

Climate statistics

	Month												
	J	F	M	A	M	J	J	A	S	O	N	D	Year
Clearness index *k* (daily average) (dimensionless)	0.428	0.522	0.570	0.619	0.661	0.675	0.738	0.701	0.685	0.619	0.495	0.441	0.596
Dry bulb daytime average) °C	0.1	3.2	6.6	11.0	15.9	20.4	25.8	24.9	19.9	13.1	6.0	1.0	12.3
(daily average) °C	−0.9	1.9	5.0	9.1	14.4	19.0	24.1	22.6	17.2	10.7	4.2	−0.1	10.6
Wind speed (daily average) m/s	3.8	4.0	4.5	4.5	4.2	4.0	3.7	3.5	3.6	3.7	3.8	3.7	3.9

Insolation on collector, kWh/m²·d

Collector type	Collector tilt	Month												
		J	F	M	A	M	J	J	A	S	O	N	D	Year
Flat plate—south facing	Horizontal	1.53	2.65	4.11	5.76	7.18	7.77	8.24	6.93	5.48	3.59	1.98	1.38	4.72
	30 degrees	2.53	3.80	5.01	6.18	7.04	7.32	7.91	7.17	6.40	4.93	3.18	2.47	5.34
	35 degrees	2.65	3.93	5.06	6.13	6.88	7.11	7.71	7.08	6.43	5.06	3.32	2.60	5.34
	40 degrees	2.76	4.03	5.09	6.05	6.69	6.86	7.46	6.94	6.43	5.16	3.44	2.72	5.31
	45 degrees	2.84	4.10	5.09	5.94	6.46	6.58	7.18	6.77	6.38	5.23	3.53	2.82	5.25
	50 degrees	2.91	4.15	5.06	5.79	6.20	6.27	6.86	6.56	6.30	5.26	3.61	2.90	5.16
	55 degrees	2.96	4.17	5.00	5.62	5.91	5.92	6.51	6.32	6.18	5.26	3.66	2.96	5.04
	60 degrees	2.99	4.17	4.91	5.41	5.59	5.56	6.12	6.04	6.03	5.23	3.69	3.01	4.90
	Vertical	2.77	3.61	3.81	3.63	3.20	2.90	3.31	3.79	4.42	4.39	3.36	2.85	3.50

Table C-3 Climatic Data for Boston MA Latitude 42.3°; Longitude 71.1

Climate statistics

	Month												
	J	F	M	A	M	J	J	A	S	O	N	D	Year
Clearness index k (daily average) (dimensionless)	0.396	0.425	0.435	0.445	0.482	0.498	0.493	0.471	0.488	0.469	0.377	0.381	0.447
Dry bulb (daytime average), °C	−1.1	−0.3	3.5	9.3	15.1	20.4	23.5	22.6	18.8	13.7	7.8	1.3	11.4
(daily average), °C	−1.7	−1.0	2.7	8.3	14.2	19.6	22.7	21.5	17.7	12.6	7.0	0.6	10.5
Wind speed (daily average), m/s	6.8	6.6	6.7	6.5	6.0	5.5	5.1	5.1	5.2	5.6	6.1	6.5	5.9

Insolation on collector, kWh/m²·d

Collector type	Collector tilt	Month												
		J	F	M	A	M	J	J	A	S	O	N	D	Year
Flat plate—south facing	Horizontal	1.50	2.24	3.21	4.18	5.26	5.73	5.52	4.69	3.97	2.81	1.59	1.27	3.50
	25 degrees	2.26	2.94	3.68	4.37	5.19	5.51	5.36	4.78	4.42	3.53	2.25	2.01	3.86
	30 degrees	2.37	3.04	3.73	4.35	5.09	5.37	5.25	4.73	4.45	3.63	2.35	2.13	3.88
	35 degrees	2.48	3.12	3.75	4.30	4.97	5.22	5.11	4.66	4.45	3.70	2.43	2.23	3.87
	40 degrees	2.56	3.18	3.75	4.24	4.83	5.04	4.95	4.56	4.42	3.75	2.50	2.32	3.85
	45 degrees	2.63	3.22	3.73	4.15	4.67	4.83	4.76	4.44	4.38	3.78	2.56	2.40	3.80
	50 degrees	2.69	3.25	3.70	4.04	4.48	4.61	4.56	4.30	4.31	3.79	2.59	2.46	3.73
	55 degrees	2.73	3.25	3.64	3.91	4.27	4.36	4.33	4.14	4.21	3.77	2.62	2.50	3.65
	Vertical	2.51	2.76	2.73	2.54	2.36	2.20	2.28	2.52	2.99	3.07	2.34	2.36	2.55

Table C-4 Climatic Data for Caribou ME Latitude 46.9°; Longitude 63.0°

Climate statistics

	Month												
	J	*F*	*M*	*A*	*M*	*J*	*J*	*A*	*S*	*O*	*N*	*D*	*Year*
Clearness index *k* (daily average) (dimensionless)	0.438	0.504	0.528	0.494	0.462	0.482	0.500	0.489	0.456	0.411	0.335	0.380	0.456
Dry bulb (daytime average), °C	−10.7	−9.2	−3.1	4.1	10.9	17.1	19.7	18.5	14.1	7.6	0.6	−7.9	5.2
(daily average), °C	−11.4	−10.1	−4.1	2.8	10.1	16.1	18.6	17.0	12.5	6.3	−0.2	−8.5	4.2
Wind speed (daily average), m/s	5.2	5.2	5.5	5.1	5.1	4.7	4.3	4.2	4.5	4.8	4.9	5.1	4.9

Insolation on collector, kWh/m²·d

Collector type	Collector tilt	Month												
		J	*F*	*M*	*A*	*M*	*J*	*J*	*A*	*S*	*O*	*N*	*D*	*Year*
Flat plate—south facing	Horizontal	1.32	2.28	3.57	4.46	4.98	5.54	5.56	4.73	3.48	2.17	1.16	0.98	3.35
	30 degrees	2.40	3.44	4.44	4.79	4.90	5.29	5.39	4.90	3.99	2.90	1.83	1.88	3.85
	35 degrees	2.54	3.57	4.51	4.77	4.80	5.15	5.27	4.84	4.00	2.97	1.91	1.99	3.86
	40 degrees	2.65	3.68	4.55	4.71	4.68	4.99	5.12	4.75	4.00	3.02	1.98	2.09	3.85
	45 degrees	2.76	3.76	4.56	4.63	4.54	4.81	4.94	4.65	3.97	3.06	2.04	2.18	3.82
	50 degrees	2.84	3.82	4.54	4.53	4.38	4.60	4.75	4.51	3.92	3.07	2.08	2.26	3.78
	55 degrees	2.90	3.86	4.50	4.41	4.19	4.38	4.53	4.36	3.85	3.07	2.11	2.31	3.71
	60 degrees	2.95	3.87	4.44	4.26	3.99	4.14	4.29	4.19	3.76	3.06	2.13	2.36	3.62
	Vertical	2.81	3.44	3.53	2.97	2.46	2.38	2.54	2.78	2.84	2.59	1.96	2.28	2.71

Table C-5 Climatic Data for Casper WY Latitude 42.9°; Longitude 106.3°

Climate statistics

	Month												
	J	F	M	A	M	J	J	A	S	O	N	D	Year
Clearness index k (daily average) (dimensionless)	0.585	0.617	0.622	0.622	0.638	0.685	0.715	0.707	0.683	0.652	0.587	0.579	0.641
Dry bulb (daytime average), °C	−2.8	−0.8	2.1	7.9	13.3	19.1	23.7	23.9	17.6	11.1	2.8	−1.5	9.7
(daily average), °C	−4.4	−2.7	−0.2	5.3	11.4	17.2	21.8	20.8	14.3	8.2	0.8	−3.0	7.4
Wind speed (daily average), m/s	7.6	6.9	6.4	5.9	5.3	5.1	4.5	4.7	5.0	5.5	6.5	7.3	5.9

Insolation on collector, kWh/m²·d

Collector type	Collector tilt	Month												
		J	F	M	A	M	J	J	A	S	O	N	D	Year
Flat plate—south facing	Horizontal	2.16	3.20	4.55	5.83	6.95	7.89	8.00	7.02	5.52	3.84	2.41	1.87	4.94
	30 degrees	3.76	4.69	5.56	6.22	6.78	7.41	7.65	7.24	6.41	5.28	3.95	3.50	5.71
	35 degrees	3.95	4.85	5.63	6.17	6.63	7.19	7.45	7.14	6.44	5.41	4.14	3.70	5.73
	40 degrees	4.12	4.97	5.66	6.09	6.44	6.94	7.21	7.00	6.43	5.52	4.29	3.88	5.71
	45 degrees	4.26	5.07	5.66	5.97	6.22	6.65	6.93	6.82	6.38	5.59	4.42	4.03	5.67
	50 degrees	4.37	5.14	5.62	5.82	5.97	6.32	6.62	6.61	6.29	5.62	4.51	4.16	5.59
	55 degrees	4.46	5.17	5.55	5.64	5.68	5.97	6.27	6.36	6.17	5.62	4.58	4.25	5.48
	60 degrees	4.51	5.17	5.45	5.43	5.37	5.59	5.90	6.07	6.01	5.59	4.62	4.32	5.34
	Vertical	4.21	4.49	4.21	3.62	3.06	2.88	3.16	3.78	4.39	4.67	4.21	4.11	3.90

Table C-6 Climatic Data for Charleston SC   Latitude 32.9°;   Longitude 79.9°

Climate statistics

	Month												
	J	*F*	*M*	*A*	*M*	*J*	*J*	*A*	*S*	*O*	*N*	*D*	*Year*
Clearness index *k* (daily average) (dimensionless)	0.437	0.467	0.497	0.545	0.532	0.510	0.507	0.482	0.486	0.514	0.512	0.460	0.496
Dry bulb (daytime average), °C	11.1	11.7	15.2	19.7	23.9	26.3	27.7	27.4	25.2	20.7	15.9	12.7	19.8
(daily average), °C	8.8	9.9	13.3	17.8	22.0	24.7	26.2	25.9	23.5	18.3	13.3	9.7	17.8
Wind speed (daily average), m/s	4.2	4.6	4.7	4.6	4.1	3.9	3.6	3.4	3.7	3.7	3.7	3.9	4.0

Insolation on collector, kWh/m²·d

Collector type	Collector tilt	Month												
		J	*F*	*M*	*A*	*M*	*J*	*J*	*A*	*S*	*O*	*N*	*D*	*Year*
Flat plate—south facing	Horizontal	2.35	3.14	4.22	5.46	5.87	5.82	5.67	5.00	4.40	3.76	2.95	2.27	4.24
	20 degrees	2.99	3.71	4.60	5.57	5.69	5.53	5.45	4.98	4.64	4.33	3.72	3.01	4.52
	25 degrees	3.11	3.81	4.64	5.53	5.58	5.39	5.32	4.91	4.65	4.42	3.87	3.15	4.53
	30 degrees	3.22	3.88	4.65	5.46	5.43	5.22	5.17	4.83	4.63	4.49	4.00	3.28	4.52
	35 degrees	3.30	3.94	4.64	5.36	5.26	5.03	4.99	4.71	4.59	4.53	4.10	3.39	4.49
	40 degrees	3.37	3.97	4.61	5.24	5.07	4.81	4.79	4.58	4.52	4.55	4.17	3.47	4.43
	45 degrees	3.42	3.98	4.55	5.08	4.85	4.57	4.57	4.42	4.43	4.54	4.23	3.54	4.35
	50 degrees	3.45	3.97	4.46	4.90	4.61	4.31	4.33	4.24	4.32	4.50	4.25	3.58	4.24
	Vertical	2.93	3.09	3.01	2.71	1.99	1.58	1.74	2.18	2.73	3.34	3.55	3.15	2.66

Table C-7 Climatic Data for Chattanooga TN Latitude 35.0°; Longitude 85.3°

Climate statistics

	Month												
	J	F	M	A	M	J	J	A	S	O	N	D	Year
Clearness index *k* (daily average) (dimensionless)	0.396	0.423	0.449	0.493	0.495	0.505	0.489	0.499	0.475	0.497	0.451	0.400	0.465
Dry bulb													
(daytime average), °C	5.3	6.7	11.0	17.4	21.8	25.3	26.9	26.6	23.6	17.3	10.9	6.9	16.7
(daily average), °C	3.9	5.6	9.7	15.5	19.7	23.6	25.2	24.8	21.7	15.1	8.8	4.9	14.9
Wind speed (daily average), m/s	3.2	3.4	3.6	3.5	2.8	2.4	2.3	2.1	2.2	2.2	2.8	3.0	2.8

Insolation on collector, kWh/m²·d

Collector type	Collector tilt	Month												
		J	F	M	A	M	J	J	A	S	O	N	D	Year
Flat plate—south facing	Horizontal	1.99	2.71	3.71	4.89	5.46	5.78	5.47	5.14	4.21	3.50	2.44	1.83	3.93
	20 degrees	2.55	3.21	4.05	5.01	5.34	5.53	5.29	5.16	4.48	4.07	3.09	2.43	4.19
	25 degrees	2.66	3.30	4.09	4.98	5.24	5.40	5.18	5.10	4.49	4.16	3.22	2.55	4.20
	30 degrees	2.76	3.37	4.11	4.92	5.12	5.24	5.04	5.02	4.48	4.23	3.32	2.66	4.19
	35 degrees	2.84	3.42	4.10	4.84	4.97	5.06	4.88	4.91	4.45	4.28	3.41	2.75	4.16
	40 degrees	2.90	3.46	4.08	4.74	4.80	4.85	4.70	4.78	4.40	4.30	3.48	2.82	4.11
	45 degrees	2.95	3.47	4.03	4.61	4.60	4.62	4.49	4.62	4.32	4.30	3.53	2.88	4.04
	50 degrees	2.97	3.46	3.96	4.46	4.38	4.37	4.26	4.44	4.22	4.28	3.55	2.92	3.94
	Vertical	2.57	2.74	2.73	2.56	2.02	1.74	1.84	2.35	2.72	3.24	3.01	2.60	2.51

Table C-8 Climatic Data for Cleveland OH Latitude 41.4°; Longitude 81.7°

Climate statistics

	Month												
	J	*F*	*M*	*A*	*M*	*J*	*J*	*A*	*S*	*O*	*N*	*D*	*Year*
Clearness index *k* (daily average) (dimensionless)	0.311	0.350	0.388	0.449	0.485	0.505	0.515	0.499	0.474	0.447	0.337	0.287	0.421
Dry bulb (daytime average), °C	−2.3	−1.2	3.2	10.4	16.2	21.8	23.8	23.2	19.5	13.6	6.8	0.4	11.3
(daily average), °C	−3.1	−1.8	2.4	9.2	14.8	20.3	22.3	21.4	17.8	11.9	5.6	−0.3	10.1
Wind speed (daily average), m/s	5.6	5.5	5.5	5.3	4.7	4.2	3.9	3.8	4.0	4.5	5.3	5.5	4.8

Insolation on collector, kWh/m²·d

Collector type	Collector tilt	Month												
		J	*F*	*M*	*A*	*M*	*J*	*J*	*A*	*S*	*O*	*N*	*D*	*Year*
Flat plate—south facing	Horizontal	1.23	1.90	2.91	4.26	5.30	5.81	5.77	4.99	3.91	2.74	1.47	1.00	3.44
	25 degrees	1.73	2.39	3.28	4.44	5.21	5.57	5.59	5.09	4.31	3.39	2.01	1.48	3.71
	30 degrees	1.81	2.46	3.31	4.41	5.11	5.43	5.47	5.03	4.33	3.47	2.09	1.55	3.71
	35 degrees	1.88	2.51	3.32	4.36	4.99	5.27	5.32	4.95	4.33	3.53	2.16	1.61	3.69
	40 degrees	1.93	2.55	3.32	4.29	4.84	5.08	5.14	4.84	4.30	3.57	2.21	1.67	3.65
	45 degrees	1.97	2.57	3.29	4.19	4.67	4.87	4.94	4.71	4.25	3.59	2.25	1.71	3.59
	50 degrees	2.01	2.58	3.25	4.08	4.48	4.64	4.72	4.55	4.18	3.60	2.28	1.75	3.51
	55 degrees	2.02	2.57	3.20	3.95	4.27	4.39	4.48	4.38	4.08	3.57	2.30	1.77	3.42
	Vertical	1.82	2.14	2.37	2.54	2.33	2.16	2.31	2.63	2.87	2.83	2.02	1.63	2.31

Table C-9 Climatic Data for El Paso TX Latitude 31.8°; Longitude 106.4°

Climate statistics

	Month												
	J	F	M	A	M	J	J	A	S	O	N	D	Year
Clearness index *k* (daily average) (dimensionless)	0.639	0.678	0.700	0.739	0.743	0.744	0.692	0.692	0.686	0.693	0.662	0.634	0.692
Dry bulb													
(daytime average), °C	9.2	11.6	15.7	20.7	25.7	30.3	30.1	29.0	26.1	20.8	13.7	9.2	20.2
(daily average), °C	6.9	9.4	13.6	18.4	23.2	27.8	27.9	26.8	23.9	18.1	11.2	6.9	17.9
Wind speed (daily average), m/s	4.0	4.5	5.1	5.3	4.8	4.4	3.9	3.7	3.6	3.6	3.7	3.7	4.2

Insolation on collector, kWh/m²·d

Collector type	Collector tilt	Month												
		J	F	M	A	M	J	J	A	S	O	N	D	Year
Flat plate—south facing	Horizontal	3.55	4.67	6.02	7.46	8.20	8.46	7.73	7.21	6.27	5.17	3.92	3.25	5.99
	15 degrees	4.44	5.47	6.55	7.63	8.04	8.14	7.51	7.24	6.63	5.88	4.81	4.17	6.38
	20 degrees	4.70	5.68	6.66	7.62	7.91	7.96	7.36	7.18	6.69	6.07	5.06	4.44	6.44
	25 degrees	4.92	5.87	6.74	7.56	7.73	7.73	7.17	7.08	6.71	6.21	5.28	4.68	6.47
	30 degrees	5.12	6.01	6.78	7.47	7.52	7.45	6.94	6.94	6.70	6.32	5.46	4.89	6.47
	35 degrees	5.28	6.12	6.77	7.33	7.26	7.15	6.68	6.77	6.64	6.40	5.61	5.06	6.42
	40 degrees	5.41	6.19	6.73	7.14	6.96	6.80	6.39	6.56	6.55	6.43	5.73	5.21	6.34
	45 degrees	5.51	6.23	6.65	6.92	6.63	6.42	6.06	6.31	6.42	6.43	5.81	5.32	6.23
	Vertical	4.79	4.87	4.33	3.48	2.42	1.88	2.02	2.85	3.81	4.72	4.90	4.78	3.73

Table C-10 Climatic Data for El Toro CA   Latitude 33.7°;   Longitude 117.7°

Climate statistics

	Month												
	J	F	M	A	M	J	J	A	S	O	N	D	Year
Clearness index *k* (daily average) (dimensionless)	0.570	0.590	0.604	0.609	0.592	0.606	0.666	0.657	0.610	0.594	0.575	0.570	0.604
Dry bulb (daytime average), °C	14.6	15.3	15.3	16.7	18.3	19.7	22.9	23.7	22.7	20.6	17.6	15.3	18.6
(daily average), °C	12.2	12.8	13.1	14.6	16.3	18.4	20.9	21.4	20.4	18.1	15.2	12.8	16.3
Wind speed (daily average), m/s	2.5	2.5	2.5	2.5	2.5	2.4	2.3	2.3	2.1	2.2	2.3	2.4	2.4
Heating degree days (C-day/mo, base 24C)	194.7	156.8	162.6	118.1	72.1	25.1	1.5	0.4	7.5	41.1	105.3	175.3	1060.4

Insolation on collector, kWh/m²·d

Collector type	Collector tilt	Month												
		J	F	M	A	M	J	J	A	S	O	N	D	Year
Flat plate—south facing	Horizontal	2.99	3.90	5.08	6.08	6.53	6.92	7.46	6.80	5.48	4.28	3.24	2.74	5.12
	20 degrees	3.98	4.75	5.63	6.24	6.35	6.58	7.15	6.82	5.87	5.02	4.18	3.78	5.53
	25 degrees	4.17	4.91	5.69	6.20	6.22	6.41	6.98	6.74	5.90	5.14	4.37	3.99	5.56
	30 degrees	4.34	5.03	5.73	6.13	6.06	6.20	6.78	6.62	5.89	5.24	4.52	4.17	5.56
	35 degrees	4.49	5.13	5.73	6.03	5.88	5.97	6.54	6.47	5.85	5.30	4.65	4.33	5.53
	40 degrees	4.60	5.19	5.70	5.89	5.66	5.70	6.27	6.28	5.78	5.33	4.75	4.46	5.47
	45 degrees	4.69	5.23	5.64	5.72	5.41	5.41	5.96	6.06	5.67	5.33	4.83	4.57	5.38
	50 degrees	4.75	5.23	5.55	5.52	5.14	5.10	5.63	5.81	5.54	5.30	4.87	4.64	5.26
	Vertical	4.14	4.14	3.76	3.04	2.21	1.83	2.16	2.88	3.48	3.98	4.12	4.16	3.32

Table C-11 Climatic Data for Fargo ND Latitude 46.9°; Longitude 96.9°

Climate statistics

	Month												
	J	F	M	A	M	J	J	A	S	O	N	D	Year
Clearness index k (daily average) (dimensionless)	0.434	0.491	0.512	0.516	0.538	0.547	0.602	0.595	0.540	0.522	0.419	0.414	0.511
Dry bulb													
(daytime average), °C	−13.7	−9.7	−2.6	7.0	14.8	20.5	23.4	22.7	16.3	10.5	−0.2	−9.2	6.7
(daily average), °C	−14.7	−10.9	−3.9	5.7	13.2	18.9	21.7	20.9	14.3	8.2	−1.7	−10.2	5.1
Wind speed (daily average), m/s	5.8	5.7	6.0	6.5	5.9	5.4	4.8	5.1	5.5	5.9	6.0	5.7	5.7

Insolation on collector, kWh/m²·d

Collector type	Collector tilt	Month												
		J	F	M	A	M	J	J	A	S	O	N	D	Year
Flat plate—south facing	Horizontal	1.31	2.23	3.46	4.66	5.79	6.29	6.69	5.76	4.11	2.76	1.44	1.06	3.80
	30 degrees	2.37	3.34	4.28	5.02	5.73	6.01	6.51	6.03	4.81	3.85	2.40	2.08	4.38
	35 degrees	2.51	3.46	4.35	4.99	5.62	5.85	6.36	5.97	4.84	3.96	2.52	2.21	4.39
	40 degrees	2.62	3.57	4.38	4.94	5.48	5.67	6.18	5.87	4.85	4.05	2.62	2.33	4.38
	45 degrees	2.72	3.65	4.39	4.86	5.31	5.46	5.97	5.74	4.82	4.11	2.70	2.43	4.35
	50 degrees	2.80	3.70	4.38	4.75	5.12	5.23	5.73	5.58	4.77	4.15	2.77	2.51	4.29
	55 degrees	2.87	3.74	4.34	4.62	4.90	4.97	5.46	5.39	4.69	4.16	2.82	2.58	4.21
	60 degrees	2.91	3.75	4.27	4.47	4.66	4.69	5.17	5.18	4.59	4.15	2.85	2.63	4.11
	Vertical	2.77	3.33	3.39	3.11	2.84	2.65	3.00	3.40	3.46	3.55	2.65	2.56	3.06

Table C-12 Climatic Data for Grand Junction CO Latitude 39.1°; Longitude 108.6°

Climate statistics

	Month												
	J	F	M	A	M	J	J	A	S	O	N	D	Year
Clearness index k (daily average) (dimensionless)	0.577	0.611	0.630	0.650	0.684	0.713	0.694	0.679	0.683	0.657	0.610	0.594	0.649
Dry bulb (daytime average), °C	−2.0	2.5	7.5	13.2	18.8	24.1	27.4	26.8	21.9	14.9	6.0	−0.3	13.4
(daily average), °C	−3.9	0.5	5.2	10.7	16.9	22.5	26.0	24.1	19.1	12.3	3.8	−2.3	11.2
Wind speed (daily average), m/s	2.5	3.1	3.8	4.3	4.3	4.4	4.2	4.0	4.0	3.5	3.0	2.6	3.7

Insolation on collector, kWh/m²·d

Collector type	Collector tilt	Month												
		J	F	M	A	M	J	J	A	S	O	N	D	Year
Flat plate—south facing	Horizontal	2.50	3.53	4.90	6.27	7.51	8.20	7.78	6.88	5.79	4.24	2.90	2.31	5.23
	25 degrees	3.80	4.73	5.72	6.59	7.33	7.76	7.47	7.01	6.48	5.42	4.24	3.72	5.86
	30 degrees	4.00	4.89	5.80	6.55	7.18	7.54	7.29	6.93	6.52	5.56	4.45	3.94	5.89
	35 degrees	4.17	5.03	5.84	6.47	6.99	7.29	7.07	6.81	6.52	5.68	4.62	4.13	5.89
	40 degrees	4.32	5.13	5.85	6.36	6.77	7.00	6.81	6.65	6.48	5.76	4.76	4.30	5.85
	45 degrees	4.44	5.20	5.82	6.21	6.51	6.68	6.52	6.45	6.41	5.81	4.88	4.44	5.78
	50 degrees	4.53	5.24	5.76	6.03	6.21	6.32	6.20	6.22	6.29	5.81	4.96	4.55	5.68
	55 degrees	4.59	5.25	5.67	5.82	5.89	5.94	5.85	5.96	6.14	5.79	5.01	4.63	5.54
	Vertical	4.18	4.40	4.15	3.57	2.94	2.61	2.74	3.37	4.20	4.64	4.46	4.33	3.79

Table C-13 Climatic Data for Kansas City MO Latitude 39.1°; Longitude 94.6°

Climate statistics

	Month												
	J	F	M	A	M	J	J	A	S	O	N	D	Year
Clearness index k (daily average) (dimensionless)	0.472	0.488	0.488	0.515	0.538	0.570	0.592	0.580	0.540	0.534	0.490	0.456	0.522
Dry bulb													
(daytime average), °C	−0.7	2.6	7.4	14.7	20.2	25.2	27.8	27.1	22.6	17.3	8.8	2.1	14.6
(daily average), °C	−1.8	1.4	5.9	13.5	19.1	24.1	26.7	25.7	21.0	15.2	7.0	0.9	13.2
Wind speed (daily average), m/s	4.5	4.6	5.1	5.1	4.6	4.4	4.0	4.0	4.0	4.2	4.5	4.5	4.5

Insolation on collector, kWh/m²·d

Collector type	Collector tilt	Month												
		J	F	M	A	M	J	J	A	S	O	N	D	Year
Flat plate—south facing	Horizontal	2.04	2.82	3.79	4.97	5.91	6.56	6.63	5.88	4.58	3.45	2.33	1.77	4.23
	25 degrees	3.00	3.65	4.32	5.16	5.76	6.22	6.37	5.96	5.04	4.29	3.28	2.71	4.65
	30 degrees	3.14	3.77	4.36	5.13	5.64	6.05	6.22	5.88	5.06	4.39	3.43	2.86	4.67
	35 degrees	3.27	3.85	4.38	5.06	5.50	5.86	6.03	5.78	5.05	4.47	3.55	2.99	4.65
	40 degrees	3.37	3.92	4.38	4.97	5.32	5.63	5.82	5.64	5.01	4.52	3.64	3.10	4.61
	45 degrees	3.46	3.96	4.35	4.86	5.12	5.38	5.58	5.48	4.95	4.54	3.72	3.19	4.55
	50 degrees	3.52	3.98	4.29	4.71	4.90	5.11	5.31	5.28	4.86	4.54	3.77	3.25	4.46
	55 degrees	3.56	3.98	4.21	4.55	4.65	4.81	5.02	5.06	4.74	4.51	3.80	3.30	4.35
	Vertical	3.21	3.30	3.08	2.84	2.41	2.21	2.42	2.91	3.26	3.59	3.35	3.05	2.97

Table C-14 Climatic Data for Kingsville TX Latitude 27.5°; Longitude 97.9°

Climate statistics

	Month												
	J	F	M	A	M	J	J	A	S	O	N	D	Year
Clearness index *k* (daily average) (dimensionless)	0.461	0.489	0.502	0.511	0.535	0.572	0.601	0.577	0.541	0.549	0.495	0.459	0.524
Dry bulb (daytime average), °C	15.4	17.5	20.8	24.2	27.1	29.5	30.7	30.4	28.6	25.4	20.1	16.4	23.8
(daily average), °C	13.6	15.4	18.8	22.7	25.4	27.8	28.9	28.8	26.8	22.8	17.9	14.5	21.9
Wind speed (daily average), m/s	4.0	4.4	4.7	5.2	4.8	4.6	4.4	3.8	3.4	3.4	3.8	3.8	4.2

Insolation on collector, kWh/m²·d

Collector type	Collector tilt	Month												
		J	F	M	A	M	J	J	A	S	O	N	D	Year
Flat plate—south facing	Horizontal	2.88	3.66	4.53	5.25	5.88	6.42	6.66	6.06	5.13	4.38	3.26	2.68	4.73
	15 degrees	3.37	4.09	4.77	5.27	5.71	6.13	6.40	6.00	5.30	4.82	3.78	3.20	4.91
	20 degrees	3.51	4.19	4.81	5.23	5.60	5.98	6.25	5.93	5.32	4.92	3.92	3.35	4.92
	25 degrees	3.63	4.28	4.83	5.17	5.46	5.79	6.07	5.83	5.30	5.00	4.04	3.47	4.91
	30 degrees	3.72	4.34	4.82	5.08	5.30	5.58	5.86	5.69	5.26	5.04	4.13	3.58	4.87
	35 degrees	3.80	4.38	4.78	4.97	5.10	5.33	5.62	5.53	5.19	5.06	4.21	3.67	4.81
	40 degrees	3.85	4.39	4.72	4.83	4.89	5.07	5.36	5.34	5.09	5.05	4.25	3.73	4.71
	45 degrees	3.88	4.38	4.64	4.66	4.65	4.77	5.06	5.12	4.97	5.01	4.27	3.77	4.60
	Vertical	3.15	3.21	2.89	2.29	1.57	1.12	1.41	2.17	2.83	3.48	3.38	3.16	2.55

Table C-15 Climatic Data for Norfolk VA Latitude 36.9°; Longitude 76.3°

Climate statistics

	Month												
	J	*F*	*M*	*A*	*M*	*J*	*J*	*A*	*S*	*O*	*N*	*D*	*Year*
Clearness index *k* (daily average) (dimensionless)	0.455	0.480	0.502	0.540	0.541	0.550	0.521	0.518	0.506	0.505	0.501	0.462	0.507
Dry bulb (daytime average), °C	6.1	6.5	10.2	15.9	20.5	24.8	26.8	26.3	23.4	17.7	12.4	7.7	16.6
Dry bulb (daily average), °C	4.6	5.4	8.9	14.4	19.0	23.4	25.4	24.8	21.9	16.2	11.0	6.3	15.1
Wind speed (daily average), m/s	5.2	5.4	5.5	5.3	4.6	4.2	3.9	3.9	4.3	4.6	4.7	5.0	4.7

Insolation on collector, kWh/m²·d

Collector type	Collector tilt	Month												
		J	*F*	*M*	*A*	*M*	*J*	*J*	*A*	*S*	*O*	*N*	*D*	*Year*
Flat plate—south facing	Horizontal	2.14	2.94	4.04	5.29	5.95	6.31	5.85	5.30	4.40	3.42	2.56	1.97	4.18
	20 degrees	2.87	3.59	4.49	5.47	5.85	6.07	5.68	5.35	4.73	4.03	3.36	2.74	4.52
	25 degrees	3.01	3.71	4.55	5.45	5.75	5.93	5.57	5.30	4.76	4.14	3.52	2.90	4.55
	30 degrees	3.14	3.81	4.58	5.40	5.62	5.76	5.43	5.23	4.77	4.22	3.65	3.04	4.56
	35 degrees	3.25	3.88	4.59	5.33	5.47	5.57	5.26	5.12	4.74	4.28	3.77	3.17	4.54
	40 degrees	3.34	3.94	4.58	5.22	5.29	5.35	5.07	4.99	4.69	4.31	3.86	3.27	4.49
	45 degrees	3.41	3.97	4.54	5.09	5.08	5.10	4.85	4.84	4.62	4.32	3.93	3.35	4.42
	50 degrees	3.45	3.98	4.47	4.93	4.84	4.83	4.61	4.66	4.52	4.30	3.97	3.41	4.33
	Vertical	3.07	3.22	3.14	2.87	2.28	1.99	2.05	2.52	2.97	3.32	3.45	3.13	2.83

Table C-16 Climatic Data for Oakland CA Latitude 37.7°; Longitude 122.3°

Climate statistics

	Month												
	J	F	M	A	M	J	J	A	S	O	N	D	Year
Clearness index k (daily average) (dimensionless)	0.489	0.535	0.578	0.622	0.634	0.645	0.653	0.635	0.623	0.575	0.522	0.495	0.584
Dry bulb (daytime average), °C	10.5	11.9	12.8	14.2	15.6	17.3	17.8	18.2	18.9	17.2	13.7	11.0	14.9
(daily average), °C	9.1	10.9	11.7	12.9	14.4	16.1	16.6	16.9	17.4	15.8	12.6	9.5	13.7
Wind speed (daily average), m/s	3.0	3.3	4.1	4.3	4.5	4.5	4.2	4.1	3.6	3.1	2.9	3.0	3.7

Insolation on collector, kWh/m²·d

Collector type	Collector tilt	Month												
		J	F	M	A	M	J	J	A	S	O	N	D	Year
Flat plate—south facing	Horizontal	2.23	3.21	4.59	6.06	6.98	7.41	7.33	6.48	5.37	3.82	2.59	2.04	4.84
	25 degrees	3.22	4.15	5.27	6.31	6.77	6.98	6.99	6.54	5.92	4.74	3.63	3.10	5.31
	30 degrees	3.37	4.28	5.33	6.27	6.62	6.78	6.81	6.45	5.94	4.85	3.78	3.26	5.32
	35 degrees	3.50	4.38	5.35	6.19	6.44	6.55	6.60	6.33	5.98	4.93	3.91	3.40	5.29
	40 degrees	3.60	4.45	5.35	6.07	6.22	6.28	6.36	6.17	5.88	4.99	4.01	3.52	5.24
	45 degrees	3.69	4.50	5.31	5.92	5.98	5.99	6.08	5.98	5.80	5.01	4.09	3.62	5.16
	50 degrees	3.75	4.51	5.24	5.74	5.70	5.66	5.77	5.76	5.69	5.00	4.14	3.69	5.06
	55 degrees	3.78	4.51	5.15	5.53	5.39	5.31	5.44	5.51	5.54	4.96	4.17	3.74	4.92
	Vertical	3.37	3.71	3.72	3.34	2.65	2.30	2.50	3.07	3.74	3.91	3.64	3.42	3.28

Table C-17 Climatic Data for Orlando FL Latitude 28.5°; Longitude 81.4°

Climate statistics

	Month												
	J	F	M	A	M	J	J	A	S	O	N	D	Year
Clearness index k (daily average) (dimensionless)	0.517	0.534	0.559	0.586	0.570	0.513	0.512	0.503	0.503	0.523	0.537	0.515	0.531
Dry bulb (daytime average), °C	17.0	17.9	20.5	23.8	26.7	28.1	28.7	28.4	27.8	25.0	21.0	17.9	23.6
(daily average), °C	15.1	16.0	18.5	21.8	24.6	26.3	27.0	26.9	26.2	23.1	19.0	15.9	21.7
Wind speed (daily average), m/s	4.2	4.5	4.6	4.4	4.0	3.8	3.4	3.2	3.6	4.0	3.9	3.9	3.9

Insolation on collector, kWh/m²·d

Collector type	Collector tilt	Month												
		J	F	M	A	M	J	J	A	S	O	N	D	Year
Flat plate—south facing	Horizontal	3.14	3.92	4.99	5.99	6.27	5.78	5.68	5.28	4.72	4.11	3.46	2.92	4.69
	15 degrees	3.75	4.43	5.30	6.05	6.10	5.54	5.49	5.24	4.89	4.53	4.06	3.56	4.91
	20 degrees	3.92	4.56	5.36	6.01	5.99	5.41	5.37	5.18	4.90	4.63	4.23	3.74	4.94
	25 degrees	4.07	4.67	5.39	6.95	5.85	5.26	5.23	5.10	4.89	4.70	4.37	3.90	4.95
	30 degrees	4.19	4.75	5.39	5.85	5.67	5.07	5.06	4.99	4.86	4.75	4.49	4.04	4.93
	35 degrees	4.29	4.80	5.36	5.72	5.47	4.87	4.87	4.85	4.79	4.77	4.53	4.15	4.83
	40 degrees	4.37	4.82	5.31	5.56	5.24	4.63	4.66	4.69	4.71	4.76	4.64	4.24	4.80
	45 degrees	4.41	4.82	5.22	5.37	4.98	4.38	4.42	4.51	4.60	4.73	4.67	4.30	4.70
	Vertical	3.65	3.59	3.27	2.63	1.73	1.18	1.39	2.02	2.68	3.32	3.75	3.67	2.73

Table C-18 Climatic Data for Philadelphia PA Latitude 39.9°; Longitude 75.2°

Climate statistics

	Month												
	J	F	M	A	M	J	J	A	S	O	N	D	Year
Clearness index k (daily average) (dimensionless)	0.417	0.442	0.455	0.472	0.478	0.497	0.495	0.492	0.481	0.477	0.423	0.395	0.460
Dry bulb (daytime average), °C	1.1	2.3	6.6	13.0	18.1	23.4	25.9	25.4	21.4	15.4	9.2	3.2	13.7
(daily average), °C	−0.2	1.1	5.2	11.5	16.9	22.1	24.5	23.7	19.8	13.7	7.7	2.1	12.3
Wind speed (daily average), m/s	4.6	4.9	5.1	5.0	4.3	3.9	3.6	3.6	3.8	4.0	4.2	4.5	4.3

Insolation on collector, kWh/m²·d

Collector type	Collector tilt	Month												
		J	F	M	A	M	J	J	A	S	O	N	D	Year
Flat plate—south facing	Horizontal	1.75	2.51	3.50	4.52	5.24	5.71	5.55	4.97	4.04	3.02	1.95	1.48	3.69
	25 degrees	2.54	3.22	3.97	4.69	5.11	5.44	5.34	5.03	4.43	3.73	2.72	2.25	4.04
	30 degrees	2.66	3.32	4.01	4.66	5.01	5.30	5.22	4.96	4.44	3.81	2.83	2.36	4.05
	35 degrees	2.77	3.39	4.03	4.60	4.89	5.14	5.07	4.88	4.43	3.88	2.93	2.47	4.04
	40 degrees	2.85	3.45	4.02	4.52	4.74	4.95	4.90	4.76	4.40	3.92	3.00	2.55	4.01
	45 degrees	2.92	3.49	3.99	4.42	4.57	4.74	4.71	4.63	4.34	3.94	3.06	2.63	3.95
	50 degrees	2.97	3.50	3.94	4.29	4.37	4.51	4.49	4.47	4.26	3.94	3.10	2.68	3.88
	55 degrees	3.01	3.50	3.87	4.15	4.16	4.26	4.26	4.29	4.15	3.91	3.12	2.72	3.78
	Vertical	2.71	2.91	2.85	2.62	2.23	2.04	2.15	2.54	2.88	3.12	2.75	2.51	2.61

Table C-19 Climatic Data for Phoenix AZ Latitude 33.4°; Longitude 112.1°

Climate statistics

	Month												
	J	*F*	*M*	*A*	*M*	*J*	*J*	*A*	*S*	*O*	*N*	*D*	*Year*
Clearness index *k* (daily average) (dimensionless)	0.610	0.652	0.679	0.743	0.765	0.757	0.701	0.698	0.706	0.687	0.640	0.606	0.687
Dry bulb													
(daytime average), °C	13.4	15.5	18.7	22.9	28.4	32.7	35.5	34.2	31.7	25.6	18.2	13.8	24.3
(daily average), °C	10.8	13.1	16.2	20.4	25.7	30.8	33.6	32.2	29.1	22.6	15.4	11.1	21.8
Wind speed (daily average), m/s	2.4	2.7	3.0	3.2	3.2	3.2	3.3	3.0	3.0	2.7	2.5	2.4	2.9

Insolation on collector, $kWh/m^2 \cdot d$

Collector type	Collector tilt	Month												
		J	*F*	*M*	*A*	*M*	*J*	*J*	*A*	*S*	*O*	*N*	*D*	*Year*
Flat plate—south facing	Horizontal	3.22	4.34	5.72	7.43	8.44	8.64	7.84	7.23	6.36	4.97	3.63	2.94	5.90
	20 degrees	4.32	5.33	6.38	7.65	8.19	8.18	7.51	7.25	6.85	5.90	4.74	4.07	6.37
	25 degrees	4.53	5.51	6.46	7.60	8.02	7.95	7.33	7.16	6.89	6.05	4.95	4.30	6.40
	30 degrees	4.72	5.66	6.51	7.52	7.81	7.69	7.11	7.04	6.88	6.17	5.14	4.50	6.40
	35 degrees	4.88	5.77	6.52	7.39	7.56	7.38	6.86	6.87	6.84	6.26	5.30	4.68	6.36
	40 degrees	5.01	5.85	6.49	7.22	7.26	7.04	6.57	6.67	6.76	6.30	5.42	4.82	6.28
	45 degrees	5.11	5.89	6.42	7.01	6.93	6.66	6.25	6.43	6.64	6.31	5.50	4.94	6.17
	50 degrees	5.18	5.89	6.31	6.76	6.56	6.25	5.89	6.16	6.48	6.28	5.55	5.02	6.03
	Vertical	4.52	4.68	4.26	3.63	2.66	2.10	2.21	3.01	4.03	4.72	4.72	4.51	3.75

Table C-20 Climatic Data for Rochester NY Latitude 43.1°; Longitude 77.6°

Climate statistics

	Month												
	J	F	M	A	M	J	J	A	S	O	N	D	Year
Clearness index k (daily average) (dimensionless)	0.315	0.343	0.392	0.452	0.466	0.498	0.503	0.483	0.454	0.421	0.312	0.276	0.410
Dry bulb (daytime average), °C	−3.4	−2.7	1.6	9.2	15.3	21.3	23.7	22.6	18.6	12.5	5.7	−0.8	10.3
(daily average), °C	−4.3	−3.6	0.5	7.7	13.7	19.5	21.8	20.7	16.6	10.8	4.8	−1.6	8.9
Wind speed (daily average), m/s	5.5	5.5	5.3	5.1	4.5	4.2	3.9	3.8	4.0	4.2	4.9	5.2	4.7

Insolation on collector, kWh/m²·d

Collector type	Collector tilt	Month												
		J	F	M	A	M	J	J	A	S	O	N	D	Year
Flat plate—south facing	Horizontal	1.15	1.76	2.85	4.22	5.07	5.73	5.62	4.79	3.66	2.47	1.27	0.89	3.29
	30 degrees	1.76	2.33	3.29	4.42	4.92	5.39	5.36	4.86	4.08	3.16	1.84	1.42	3.58
	35 degrees	1.84	2.38	3.31	4.37	4.81	5.24	5.23	4.79	4.08	3.22	1.91	1.48	3.56
	40 degrees	1.90	2.42	3.31	4.31	4.67	5.06	5.06	4.69	4.06	3.26	1.96	1.54	3.52
	45 degrees	1.94	2.45	3.29	4.22	4.52	4.86	4.87	4.57	4.02	3.28	2.00	1.58	3.47
	50 degrees	1.98	2.47	3.26	4.11	4.34	4.63	4.67	4.43	3.95	3.29	2.03	1.62	3.40
	55 degrees	2.01	2.46	3.20	3.98	4.14	4.39	4.44	4.26	3.87	3.27	2.04	1.64	3.31
	60 degrees	2.02	2.45	3.14	3.83	3.93	4.13	4.19	4.08	3.77	3.24	2.05	1.66	3.21
	Vertical	1.84	2.08	2.41	2.60	2.33	2.24	2.36	2.62	2.76	2.67	1.82	1.54	2.27

APPENDIX D

UNIFORM PRESENT WORTH TABLES*

Table D-1 Discount Rate = 0 Percent

Economic life	Escalation rate (percent)							
	0	4	6	8	10	12	15	20
5	5.00	5.63	5.93	6.34	6.72	7.12	7.75	8.93
8	8.00	9.58	10.49	11.49	12.58	13.78	15.79	19.80
10	10.00	12.49	13.97	15.65	17.53	19.65	23.35	31.15
12	12.00	15.63	17.88	20.50	23.52	27.03	33.35	47.50
15	15.00	20.82	24.67	29.32	34.95	41.75	54.72	86.44
20	20.00	30.97	38.99	49.42	63.00	80.70	117.81	224.03
25	25.00	43.31	58.16	78.95	108.18	149.33	244.71	566.38
30	30.00	58.33	83.80	122.35	180.94	270.29	499.96	1418.26

Source: Gupta.Y. P. and Young, S. K., *Design Handbook for Photovoltaic Power Systems,* Vol. 2, Science Applications Inc., McClean Va., 1981.

Table D-2 Discount Rate = 5 Percent

Economic life	Escalation rate (percent) 0	4	6	8	10	12	15	20
5	4.33	4.86	5.14	5.45	5.76	6.09	6.62	7.60
8	6.46	7.66	8.35	9.10	9.92	10.81	12.31	15.28
10	7.72	9.49	10.54	11.71	13.03	14.51	17.06	22.41
12	8.86	11.28	12.77	14.48	16.45	18.71	22.76	31.72
15	10.38	13.91	16.20	18.93	22.21	26.13	33.51	51.29
20	12.46	18.12	22.13	27.24	33.78	42.17	59.44	107.59
25	14.09	22.13	28.34	36.81	48.39	64.32	100.29	217.37
30	15.37	25.95	34.86	47.82	66.82	94.91	164.68	431.39

Table D-3 Discount Rate = 8 Percent

Economic life	Escalation rate (percent) 0	4	6	8	10	12	15	20
5	3.99	4.47	4.73	5.00	5.28	5.58	6.06	6.94
8	5.75	6.78	7.36	8.00	8.70	9.46	10.72	13.23
10	6.71	8.17	9.04	10.00	11.08	12.28	14.36	18.68
12	7.54	9.47	10.65	12.00	13.55	15.32	18.48	25.41
15	8.56	11.24	12.96	15.00	17.43	20.31	25.71	38.57
20	9.82	13.78	16.53	20.00	24.39	29.95	41.26	72.25
25	10.67	15.88	19.79	25.00	32.01	41.50	62.54	129.30
30	11.26	17.62	22.75	30.00	40.37	55.37	91.67	225.90

Table D-4 Discount Rate = 10 Percent

Economic life	Escalation rate (percent) 0	4	6	8	10	12	15	20
5	3.79	4.24	4.48	4.73	5.00	5.28	5.72	6.54
8	5.33	6.27	6.80	7.37	8.00	8.68	9.82	12.07
10	6.14	7.44	8.20	9.05	10.00	11.06	12.87	16.65
12	6.81	8.49	9.51	10.67	12.00	13.52	16.21	22.09
15	7.61	9.86	11.30	12.99	15.00	17.38	21.80	32.26
20	8.51	11.69	13.87	16.59	20.00	24.30	32.95	56.38
25	9.08	13.07	16.00	19.87	25.00	31.87	46.88	93.66
30	9.43	14.11	17.78	22.86	30.00	40.15	64.27	151.24

Table D-5 Discount Rate = 12 Percent

Economic life	Escalation rate (percent) 0	4	6	8	10	12	15	20
5	3.60	4.03	4.25	4.49	4.74	5.00	5.42	6.18
8	4.97	5.81	6.29	6.82	7.38	8.00	9.03	11.05
10	5.65	6.80	7.48	8.23	9.07	10.00	11.60	4.90
12	6.19	7.66	8.54	9.55	10.69	12.00	14.31	9.33
15	6.81	8.72	9.93	11.35	13.03	15.00	18.65	27.22
20	7.47	10.05	11.79	13.95	16.64	20.00	26.71	44.62
25	7.84	10.96	13.21	16.12	19.95	25.00	35.90	69.17
30	8.06	11.59	14.28	17.93	22.97	30.00	46.38	103.85

Table D-6 Discount Rate = 15 Percent

Economic life	Escalation rate (percent) 0	4	6	8	10	12	15	20
5	3.35	3.74	3.94	4.16	4.38	4.62	5.00	5.69
8	4.49	5.22	5.64	6.09	6.58	7.12	8.00	9.73
10	5.02	6.00	6.56	7.20	7.90	8.67	10.00	12.73
12	5.42	6.63	7.35	8.17	9.09	10.15	12.00	16.00
15	5.85	7.36	8.31	9.41	10.71	12.22	15.00	21.44
20	6.26	8.19	9.47	11.03	12.96	15.33	20.00	32.22
25	6.46	8.69	10.24	12.22	14.76	18.05	25.00	45.55
30	6.57	8.99	10.76	13.08	16.20	20.44	30.00	62.04

Table D-7 Discount Rate = 17 Percent

Economic life	Escalation rate (percent) 0	4	6	8	10	12	15	20
5	3.20	3.56	3.75	3.96	4.17	4.39	4.75	5.40
8	4.21	4.88	5.26	5.67	6.12	6.61	7.41	8.98
10	4.66	5.54	6.05	6.61	7.23	7.93	9.11	11.52
12	4.99	6.05	6.69	7.41	8.22	9.14	10.75	14.20
15	5.32	6.63	7.45	8.39	9.49	10.77	13.10	18.48
20	5.63	7.24	8.30	9.58	11.14	13.05	16.77	26.37
25	5.77	7.58	8.82	10.38	12.35	14.88	20.13	35.33
30	5.83	7.77	9.14	10.91	13.25	16.36	23.22	45.49

Table D-8 Discount Rate = 20 Percent

Economic life	*Escalation rate (percent)*							
	0	*4*	*6*	*8*	*10*	*12*	*18*	*20*
5	2.99	3.32	3.50	3.69	3.88	4.08	4.41	5.00
8	3.84	4.43	4.76	5.13	5.52	5.94	6.64	8.00
10	4.19	4.95	5.38	5.86	6.39	6.98	7.97	10.00
12	4.44	5.33	5.86	6.46	7.13	7.88	9.20	12.00
15	4.68	5.74	6.39	7.15	8.02	9.03	10.85	15.00
20	4.87	6.13	6.94	7.91	9.07	10.48	13.18	20.00
25	4.95	6.32	7.23	8.35	9.75	11.51	15.06	25.00
30	4.98	6.41	7.39	8.62	10.19	12.23	16.58	30.00

Table D-9 Discount Rate = 25 Percent

Economic life	*Escalation rate (percent)*							
	0	*4*	*6*	*8*	*10*	*12*	*15*	*20*
5	2.69	2.98	3.18	3.29	3.46	3.64	3.92	4.43
8	3.33	3.82	4.09	4.38	4.70	5.04	5.60	6.69
10	3.57	4.17	4.51	4.88	5.29	5.74	6.50	8.04
12	3.73	4.41	4.81	5.25	5.75	6.31	7.27	9.29
15	3.86	4.64	5.11	5.64	6.26	6.96	8.21	10.99
20	3.95	4.83	5.37	6.01	6.76	7.66	9.33	13.39
25	3.98	4.90	5.49	6.19	7.03	8.06	10.07	15.35
30	4.00	4.93	5.54	6.27	7.17	8.30	10.56	16.95

APPENDIX E

UNITS AND CONVERSION FACTORS

Table E-1 Conversion Factors

LENGTH

cm	in.	ft	mi	nm	km
1	3.937^{-1}	3.28^{-2}			
2.54	1	8.33^{-2}			
30.48	12	1			
		5280	1	8.685^{-1}	1.6093
		6080	1.1515	1	1.8531
		3281	6.214^{-1}	5.396^{-1}	1

VOLUME

cm³	ft³	liter	yd³	gallon
1	3.531^{-5}	1.0^{-3}	1.308^{6}	2.642^{-4}
2.832^{4}	1	28.32	3.704^{-2}	7.481
10^{3}	3.531^{-2}	1	1.308^{-3}	2.642^{-1}
7.646^{5}	27	7.646^{2}	1	202
3.785^{3}	1.337^{-1}	3.785	4.951^{-3}	1

PRESSURE

torr	mmHg	micron (μ Hg)	Pa	N/m²
1	1	1^{-3}	133.322	133.322
7.501^{-3}	7.501^{-3}	7.501	1	1

AREA

cm²	in²	ft²
1	0.155	1.08^{-3}
6.45	1	6.94^{-3}
929	144	1

DENSITY

g/cm³	lb/ft³
1	62.43
0.016	1

VELOCITY

cm/s	ft/s	ft/h	mi/h
1	0.0328	118.1	0.0224
30.48	1	3,600	0.6818
8.47^{-3}	2.78^{-4}	1	1.89^{-4}
44.70	1.467	5,280	1

TIME
(Based on 24-hour day, 30-day month, 12-month year)

s	min	h	day	week	mo	yr
1						
60	1					
3.6^{3}	60	1				
8.64^{4}	1.44^{3}	24	1			
6.05^{5}	1.01^{4}	168	7	1		
2.59^{6}	4.32^{4}	720	30	4.29	1	
3.11^{7}	5.18^{5}	8.64^{3}	360	51.4	12	1

Note: Exponents indicate powers of 10.

Table E-2 More Conversion Factors

HEAT AND SOLAR FLUX

kW/m^2	mW/cm^2	$cal/cm^2 \cdot s$	$Btu/ft^2 \cdot h$
1	100	0.0239	317.3
0.01	1	2.39^{-4}	3.173
41.83	4.183^{3}	1	1.327^{4}
3.15^{-4}	3.15^{-2}	7.53^{-5}	1

POWER

hp	kW	ft · lb/s	Btu/h	cal/s	MeV/s
1	0.7457	550	2,547	178.2	4.65^{15}
1.341	1	737.6	3,415	239	6.24^{15}
1.818^{-3}	1.356^{-3}	1	4.63	0.324	8.46^{12}
2.546	1.899	1,400	6,480	453.9	1.18^{16}
3.93^{-4}	2.93^{-4}	0.216	1	0.070	1.82^{12}
5.61^{-3}	4.18^{-3}	3.088	14.29	1	2.61^{13}
2.15^{-16}	1.60^{-16}	1.118^{-13}	5.47^{-13}	3.83^{-14}	1

ENERGY

kWh	Btu	ft · lb	cal	MeV	erg
1	3413	2.66^{6}	8.60^{5}	2.24^{19}	3.6^{13}
5.27^{-4}	1.8	1,400.6	453.6	1.18^{16}	1.9^{10}
2.93^{-4}	1	778.1	252	6.58^{15}	1.06^{10}
3.77^{7}	1.29^{-3}	1	0.324	8.46^{12}	1.36^{7}
1.16^{-6}	3.97^{-3}	3.087	1	2.61^{13}	4.19^{7}
4.46^{-20}	1.52^{-16}	1.18^{-13}	3.83^{-14}	1	1.60^{-6}
2.78^{-14}	9.48^{-11}	7.38^{-8}	2.39^{-8}	6.23^{5}	1

Note: Exponents indicate powers of 10.

Table E-3 Temperature Conversions

Celsius to Kelvin:	$T_K = T_C + 273.15$
Farenheit to Kelvin:	$T_K = (5/9)(T_F + 459.67)$
Rankine to Kelvin:	$T_K = (5/9)T_R$
Farenheit to Celsius:	$T_C = (5/9)\ (T_F - 32)$

°K	°C	°F	°R
0	−273	−460	0
73	−200	−328	132
173	−100	−148	312
233	−40	−40	420
273	0	32	492
373	100	212	672

ΔT

°F or °R	°C or °K
1.8	1
1	0.5556

APPENDIX F
DC ELECTRICAL EQUATIONS AND LAWS

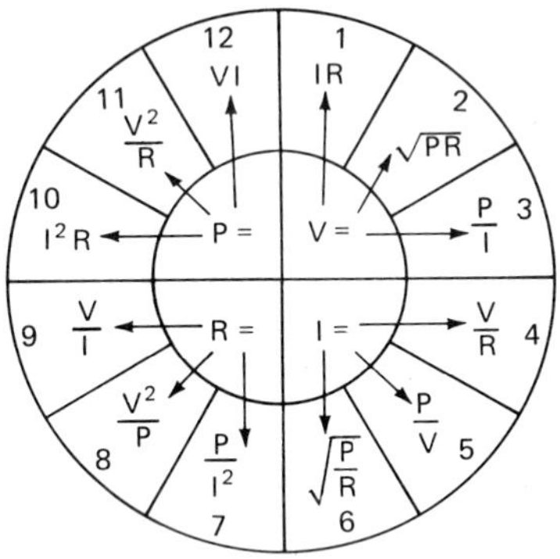

OHM's LAW: $I = \frac{V}{R}$

where I = current, A

V = voltage, V

R = resistance, Ω

POWER FORMULA: $P = I \times V$

where P = power, W

Figure F-1 **The Ohm's law and power formula memory-aid circle.**

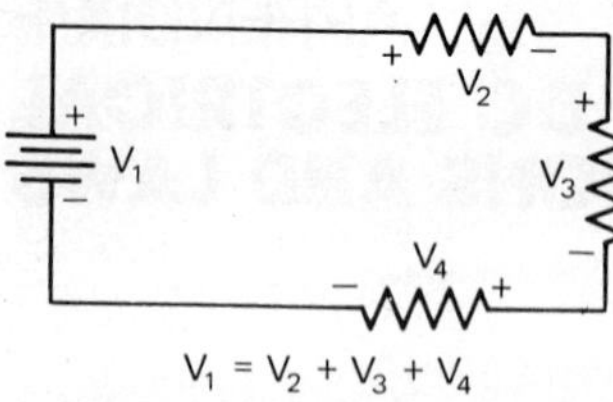

Figure F-2 **KIRCHHOFF'S VOLTAGE LAW: The algebraic sum of the potential rises and drops around a closed loop (or path) is zero.**

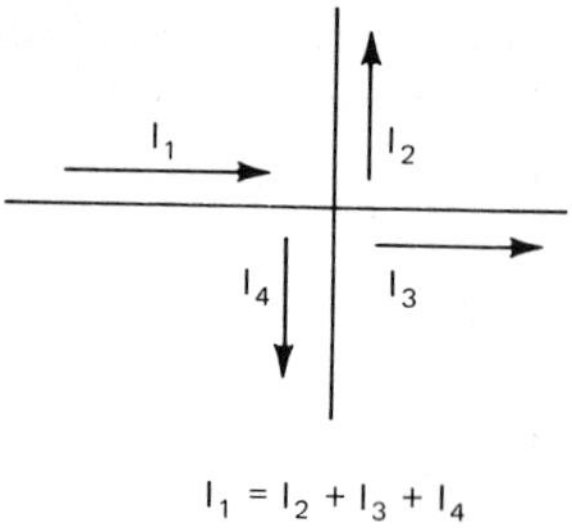

Figure F-3 **KIRCHHOFF'S CURRENT LAW: The sum of the currents entering a node must equal the sum of the currents leaving a node.**

Glossary

Air mass A ratio composed of the distance of the earth's atmosphere through which direct-beam sunlight must pass divided by the perpendicular distance traversed by the sun's rays when the sun is directly overhead at sea level.

Alternating current Electric current that periodically switches its direction (typically 60 cyles per second in the U.S. and 50 in Europe).

Altitude angle At any moment in time, the number of degrees between the sun's position and the line perpendicular to the horizon.

Ambient temperature The temperature of the air surrounding a given object, usually measured outside any building.

Amorphous A state of matter in which the atoms are arranged in a random fashion; not crystalline.

Ampere A unit of electric current equivalent to one coulomb (6.3×10^{18} electrons) passing one point each second.

Ampere-hour An energy capacity rating, used when the voltage of a source remains relatively constant, equivalent to one ampere of current flowing for one hour.

Array, photovoltaic A network of photovoltaic modules interconnected both in parallel and in series.

Azimuth angle The number of degrees between the north/south plane and the plane between the sun's position and the origin of sight; east is negative and west is positive.

Band-gap energy The energy needed to move an electron in the outer shell of an atom from the valence level to the conduction level, and thereby to create a free electron.

Blocking diode A diode placed at the positive end of a branch of photovoltaic modules in order to prevent accidental feedback of current from another branch or power-conditioning device.

Cell junction The area of contact between the P- and the N-type materials in a photovoltaic cell.

Circuit, electric A path which allows electric current to flow through some useful device as it passes from one terminal of a power source to the other.

Conduction level (or band) The energy level where electrons are not bound to any atomic nucleus and therefore are free to flow.

Current, electric The flow of electrons in a circuit.

Declination The angle between the sun and the plane of the earth's equator, as seen at solar noon from a point on the equator (varies between plus and minus 23.5 degrees depending on the time of year).

Deep-discharge battery An electric storage battery that can be repeatedly charged and discharged (to about a 30% level) without appreciably losing efficiency or capacity within its specified lifetime.

Depletion zone The small region around a photovoltaic cell's junction where all charge carriers, be they electrons or holes, have been depleted.

Diffuse radiation Sunlight that reaches the earth's surface *indirectly* because of the scattering effect of moisture, dust, etc., in the earth's atmosphere.

Direct-beam radiation Sunlight that shines *directly* from the sun to a particular surface on earth.

Direct current Electric current that flows in one uniform direction from positive to negative.

Discount rate The value of future money in terms of its present worth, usually measured as an annual percentage.

Doping A process of adding different elements to a pure material in order to alter its electrical properties and produce a semiconductor effect.

Efficiency, energy The ratio of energy output to energy input for any energy-conversion device.

Energy, electric The ability to do work by means of a difference in electric potential.

Equinoxes The two times each year when the sun's path through the earth's sky is in the plane of the equator and therefore the lengths of day and night are equal (occuring about March 21st and September 21st).

Fill factor The ratio of a photovoltaic cell's maximum power to the product of its open-circuit voltage and short-circuit current; an important characteristic of a photovoltaic cell or array which indicates its series and shunt resistance and the shape of its *I-V* curve.

Fixed flat-plate collector A solar collector that is directly exposed to the sun in a fixed position (without sunlight concentration or sun tracking).

Giga A prefix meaning multiplied by 10^9 (or 1,000,000,000).

Grid, utility The interconnecting network of the utility power system, involving power plants, transmission and distribution lines, substations, transformers, etc.

Hole A positive-charge carrier in semiconductor material created by structuring atoms in a crystal so that there is an electron vacancy.

Hot spot A defective high-resistance region in a photovoltaic array that can act as a load when current is flowing and can result in overheating, cracking, and melting of a photovoltaic module.

Incident angle The angle between a sunbeam striking a surface and the perpendicular to that surface.

Insolation The total solar radiation (direct, diffuse, and reflected) received on a surface over a given period of time.

Inverter An electric device that converts direct current into alternating current.

***I-V* curve** A graph relating the current to the voltage characteristics of a photovoltaic cell or array as the attached load is varied from infinity to zero.

Kilo A prefix meaning multiplied by 10^3 (or 1,000).

Load Any device or system of devices that uses electric energy.

Magnetic declination The angle between true north and the north measured by a compass for any given location.

Majority carriers Charge carriers in a layer of semiconductor material which are in the majority (i.e., electrons in N-type and holes in P-type material).

Maximum power A power source's greatest possible output, which can be achieved only under a specific load condition.

Mega A prefix meaning multiplied by 10^6 (or 1,000,000).

Milli A prefix meaning multiplied by 10^{-3} (or 0.001).

Minority carriers Charge carriers in a layer of semiconductor material which are in the minority (i.e., holes in N-type and electrons in P-type material).

Mismatch An incompatibility created when photovoltaic cells or modules either with different output voltages are connected in parallel or with different output currents are connected in series; this mismatch can result in power loss within an array or module.

Module, photovoltaic A subarray of solar cells that has been encapsulated into a sealed package with a transparent cover in order to protect the solar cells from environmental damage.

N-type silicon Silicon mixed with a very small quantity of a dopant element, such as phosphorus, which causes the material to have more electrons than required to complete the crystalline structure; nonetheless, no electrical imbalance is created.

NOCT Nominal operating cell temperature; it is measured under nominal operating conditions of 80 mW/cm² insolation, 20°C ambient temperature, 1 m/sec average windspeed, and open-circuit conditions.

Ohm A unit of resistance to the flow of electric current equal to the resistance of a circuit in which a potential difference of one volt produces a current of one amp.

Open circuit A break or very high resistance in an electric circuit that prevents the flow of current.

P-type silicon Silicon mixed with a very small quantity of a dopant element, such as boron, which causes the material to have fewer electrons than required to complete the crystalline structure; nonetheless, no electric imbalance is created.

Packing factor The ratio of active cell or module area to total module or array area.

Peak power The rate at which a photovoltaic cell or array generates electric energy when exposed to a "one sun" insolation of 1000 W/m² or 100 mW/cm².

Peak watt A unit of peak power output for a photovoltaic cell or array.

Photon A discrete packet of radiant energy used as the basic unit of measure for light; a quantum of electromagnetic radiation which fits the relationship $E = hv$, where E is energy, h is Planck's constant, and v is the radiation frequency.

Photovoltaic cell A single solar-cell wafer capable of converting sunlight into electricity; under a fixed illumination and normal temperatures, a silicon cell's voltage output cannot exceed 0.5 to 0.7 volts and the current and power output is proportional to cell area.

Polycrystalline cell A solar cell composed of numerous small crystals in a random order.

Power The rate at which energy is either generated or used.

Power conditioner A piece of electrical equipment (such as an inverter, a transformer, a voltage regulator, or various controls) sometimes needed to convert photovoltaic power into a more useful form.

PV Photovoltaic.

Pyranometer An instrument for measuring the total direct, diffuse, and reflected solar insolation, based on use of thermocouples to measure the temperature of a black surface.

Quad A unit of energy equal to one quadrillion (10^{15}) British thermal units or 2.93×10^{12} kilowatthours; not to be confused with the scientific unit Q which equals 10^{18} Btu, one quad equals one mQ; total U.S. energy use in 1980 was about 78 quads.

Recombination The reuniting of a hole and a free electron, as normally occurs in semiconductor material.

Reflected radiation The indirect sunlight that reaches a particular surface on earth by being reflected from another body such as the ground, a cloud, or a building.

Remote site An installation that requires electric power but does not have ready access to the electric-utility grid.

Resistance The property of a body whereby it opposes and limits the flow of current through it.

Semiconductor A type of material with electrical properties characteristic both of conductors (such as metals) and insulators (such as plastic); the important value of a semiconductor is that conduction can be limited to one direction or controlled to specific levels by external means, as achieved in diodes, transistors, and photovoltaic cells.

Short circuit A very low-resistance path between the two terminals of a power source or between two points in a circuit, usually causing current to detour away from a useful path.

Shunt or bypass diode A diode wired across the terminals of a photovoltaic module to divert string or array current in the event of shading or an open-circuit fault within the module.

Single crystal A crystalline solid in which all the atoms are arranged in a uniform and ordered fashion.

Solar constant The solar intensity that would strike a surface perpendicular to the sun's rays directly outside the earth's atmosphere (typically equal to 1353 W/m^2).

Solar fraction The fraction of a load's total energy input that is supplied by solar energy.

Solar time Time based on the sun's apparent path through the earth's sky, as distinct from Local Standard Time; solar time takes into account the eccentricities in the earth's orbit about the sun and ignores time-zone and daylight-savings considerations; solar noon is that moment during any day when the sun is at its highest altitude in the sky.

Solstices The two times each year (around June 22 and December 22) when the sun's apparent path in the daytime sky reverses its north-to-south or south-to-north direction; days and nights are either their longest or shortest at this time.

Standalone system An electric power system that is independent of the electric utility grid.

Valence band The outermost orbit of electrons in an atom or molecule.

Voltage The electromotive force that can move electrons in a circuit.

Voltage regulator An electric device that maintains the voltage of a power source at a desirable level despite changes in the load.

Watt A unit of measure of electric power equal to one joule per second.

Watthour A unit of measure of electric energy equal to the energy transferred at a rate of one watt during a period of one hour.

Index

About the Author

Matthew E. Buresch received his B.A. degree in the History of Science from Antioch University and his A.S. degree in Electronic Engineering Technology from Wentworth Institute of Technology. Since then he has worked for three years at M.I.T. Lincoln Laboratory in a research group developing photovoltaic applications at remote sites and in utility-interactive homes. His major work has involved assisting in the final checkout and turn-on of the standalone 100-kWp PV system at the Natural Bridges National Monument in Utah. He has also acted as site engineer for the utility-interactive Northeast Residential Photovoltaic Experiment Station in Massachusetts.

About the Author